Molecular Mechanisms Underpinning the Development of Obesity

Clévio Nóbrega • Raquel Rodriguez-López
Editors

Molecular Mechanisms Underpinning the Development of Obesity

Springer

Editors
Clévio Nóbrega
Center for Neurosciences and Cell Biology
University of Coimbra
Coimbra
Portugal

Raquel Rodriguez-López
Laboratory of Molecular Medicine
Clinical Analysis Service
Consorcio Hospital General de Valencia
Valencia
Spain

ISBN 978-3-319-12765-1 ISBN 978-3-319-12766-8 (eBook)
DOI 10.1007/978-3-319-12766-8

Library of Congress Control Number: 2014958583

Springer Cham Heidelberg New York Dordrecht London

Printed on acid-free paper

Springer is part of Springer Science+Business Media (www.springer.com)

Preface

On the Molecular Mechanisms Underpinning the Development of Obesity

Obesity and overweight are defined as abnormal or excessive body fat accumulation, possibly due to an imbalance of excess energy intake and reduced energy expenditure. This excessive fat accumulation can reach levels capable of affecting health, thus contributing to an increased risk of developing certain disorders.

The existence of obesity seems almost as ancient as humans, and from an evolutionary perspective it is very tempting to consider as a strategy for storing food reserves. In fact, during most time of human history, obesity could be an advantage character in times where food was scarce. The ability to store energy as fat from the least possible amount of food intake might have constituted a selective advantage. Thus, during many times in human history obesity was considered beneficial. Only in the XVIII century people start realizing that excessive body fat could have a negative impact in life quality. With the documentation of morbid complication and increased mortality associated with obesity it started to be recognized as a cause of ill health. In 1997 the World Health Organization (WHO) declare it a global epidemic and worldwide public-health crisis. Actually, the WHO considers overweight and obesity major risk factors for a number of chronic diseases, including diabetes, cardiovascular diseases and cancer. In 2010, overweight and obesity were estimated to cause more than 3.4 million deaths worldwide. Following an intense debate and controversy, in 2013 the American Medical Association recognized obesity as chronic disease in an attempt to reverse the epidemic rise of obesity. According to this decision, the diagnosis and treatment of obesity becomes a professional obligation for physicians. All these health consequences contribute to important economic costs for health systems, with estimates suggesting that obesity is responsible for approximately 1–3 % of total health expenditure in most countries (in United States the estimates point to a cost of 5–10 % of health expenditures).

Obesity was first considered a problem only of wealthy nations, however currently it has an important impact in developing countries, affecting economy, productivity and health worldwide. According to WHO the obesity levels across the world nearly doubled in the past 30 years, and the current numbers are scary. In 2008, more than 1.4 billion adults (20 years of age and older) were overweight, which corresponded to 20 % of world population. From those, almost 200 million men and 300 million women were obese. In 2012 more than 40 million children under the age of 5 were overweight or obese. In developed countries from The Organization for Economic Co-operation and Development (OECD) around 53 % of the population are overweight or obese, with 21 from the 34 countries of OECD having a prevalence above 50 % in adults. Over the past 10 years the obesity prevalence increased more than 40 % in countries like Denmark, France or Sweden and the OECD average raised from 13 % in 2000 to 18 % in 2010. In a recent study reporting obesity trends worldwide from 1980 to 2013 found an increase not only in the proportion of overweight or obese adults (both men and women), but also in children and adolescents. According to the study, the proportion of adults with a BMI of 25 kg/m^2 or greater increased in 33 years from 28.8 to 36.9 % in men, and from 29.8 to 38 % in women. In children and adolescents the prevalence of obesity and overweight has increased in developed countries (23.8 and 22.6 %, in boys and girls, respectively), but also in developing countries with an increase around 5 % from 1980 to 2013.

But how is obesity defined? As it was mentioned overweight and obesity are an excessive accumulation of body fat, which is difficult to measure. Therefore, measures based on quantitative anthropometric characteristics are used to define obesity. Currently, the body mass index (BMI), which is a simple index of weight-for-height, is most commonly used to classify overweight and obesity in adults. It is defined as a person's weight in kilograms divided by the square of his height in meters (kg/m^2). The BMI is a measure that was devised in the 19th century by Adolphe Quetelet, although only in 1972 did it become a world reference for measuring body fat. Decades of research have shown that BMI provides a good estimate for body fat, although more sophisticated and accurate measures are also being used nowadays. Currently, it is widely accepted that BMI is a reliable and easy way to access to body fat, and those are two of the most important advantages of it use as an obesity measure. Also importantly, several studies related the risk of developing health problems and risk of death with BMI. According to WHO for a healthy adult the BMI should range from 18.5 to 24.9 kg/m^2. Overweight is defined as a BMI of 25–29.9 kg/m^2, whereas obesity is defined as a BMI above 30 kg/m^2. If for adults these cutoffs are more or less conventional, the definition of children obesity based in BMI was more controversial. Depending on the age and gender, it is normal for children and adolescents to have different amounts of body fat, thus several scales based in age and gender are currently used (WHO, United States the Centers for Disease Control and Prevention, and International Obesity Task Force). Despite the great advantages and the broad use, BMI have also some limitations. For example, being an indirect measure of body fat is does not distinguish between body fat and lean body mean. Also, it is not so accurate in younger ages compared to adults and it does not take into account normal differences between gender or ethnic groups. Thus, other indirect methods were developed to measure body fat (for example waist circumference or waist-to-hip-ratio), and currently more sophisticated direct methods such as magnetic resonance imaging or dual energy X-ray absorptiometry are being used. However, despite all these methods and techniques, BMI remains the simple, cheap and most used measure of obesity.

Obesity and overweight arises in a simple way as an energy imbalance between calories consumed and calories expended. However, currently obesity is understood as a multifactorial disorder. If it is clear the effect of environment and food consumption in obesity, it is also widely accepted the effect of a genetic component in the increased risk of developing obesity. Moreover, it becomes increasingly clear the contribution of other mechanisms (like epigenetics, neurotrophic factors, microRNAs) in obesity. However, the factors underlying the development of obesity are still largely unknown. If in monogenic forms of obesity, the gene and the particular causing mutation is identified, in polygenic forms of obesity the contributions of mutations and genes are not yet completely understood. What emerges from the last decades of research is that obesity is a complex trait resulting from interplay of an obesogenic environment and a genetic background. In this book entitled: *Molecular Mechanisms Underpinning the Development of Obesity* we try to provide a state-of-the-art revision about the molecular mechanisms that could be in the basis for developing obesity, reviewing the current knowledge in areas like monogenic and polygenic obesity forms; but also providing an updated view of the emerging knowledge about epigenetics, microRNAs, and neuronal aspects that may also contribute to obesity.

Clévio Nóbrega

Contents

Contributors

Charbel Abi Khalil Department of Genetic Medicine and Department of Medicine, Weill Cornell Medical College—Qatar, Doha, Qatar

David Albuquerque Faculdade de Ciências e Tecnologia, University of Coimbra, Coimbra, Portugal

Luigi Aloe Laboratory of Cell Biology, Department of Anatomy and Histology, Medical University, Varna, Bulgaria

Haruka Amitani Department of Psychosomatic Internal Medicine, Kagoshima University Graduate School of Medical and Dental Sciences, Kagoshima, Japan

Akihiro Asakawa Department of Psychosomatic Internal Medicine, Kagoshima University Graduate School of Medical and Dental Sciences, Kagoshima, Japan

Kyle Burger Oregon Research Institute, Eugene, OR, USA

Cláudia Cavadas Center For Neuroscience and Cell Biology & Faculty of Pharmacy, University of Coimbra, Coimbra, Portugal

George N. Chaldakov Laboratory of Cell Biology, Department of Anatomy and Histology, Medical University, Varna, Bulgaria

Béatrice Dubern Nutrition and Gastroenterology Department, Armand-Trousseau Hospital, Paris, France

Institute of Cardiometabolism and Nutrition, ICAN, Pitié-Salpêtrière Hospital, Paris, France

INSERM, UMR_S U1166, Nutriomics, Pitié-Salpêtrière Hospital, Paris, France

Marco Fiore Laboratory of Cell Biology, Department of Anatomy and Histology, Medical University, Varna, Bulgaria

Emilio González-Jiménez Department of Nursing, Faculty of Health Sciences, University of Granada, Granada (Spain)

Hélène Huvenne Pediatrics Department, Saint-Vincent de Paul Hospital, Lille, France

Institute of Cardiometabolism and Nutrition, ICAN, Pitié-Salpêtrière Hospital, Paris, France

INSERM, UMR_S U1166, Nutriomics, Pitié-Salpêtrière Hospital, Paris, France

Akio Inui Department of Psychosomatic Internal Medicine, Kagoshima University Graduate School of Medical and Dental Sciences, Kagoshima, Japan

Goro Katsuura Department of Psychosomatic Internal Medicine, Kagoshima University Graduate School of Medical and Dental Sciences, Kagoshima, Japan

Namiko Kawamura Department of Psychosomatic Internal Medicine, Kagoshima University Graduate School of Medical and Dental Sciences, Kagoshima, Japan

Licínio Manco Faculdade de Ciências e Tecnologia, University of Coimbra, Coimbra, Portugal

Miyuki Nishida Department of Psychosomatic Internal Medicine, Kagoshima University Graduate School of Medical and Dental Sciences, Kagoshima, Japan

Clévio Nóbrega Center for Neurosciences and Cell Biology, University of Coimbra, Coimbra, Portugal

Luís Pereira de Almeida Center For Neuroscience and Cell Biology & Faculty of Pharmacy, University of Coimbra, Coimbra, Portugal

Raquel Rodríguez-López Laboratory of Molecular Medicine, Clinical Analysis Service, Consorcio Hospital General de Valencia, Valencia, Spain

Thorkild I. A. Sørensen Novo Nordisk Foundation Center for Basic Metabolic Research, Section on Metabolic Genetics, Faculty of Health and Medical Sciences, University of Copenhagen, Denmark, and Institute of Preventive Medicine, Bispebjerg and Frederiksberg Hospitals, The Capital Region, Copenhagen, Denmark

Lígia Sousa-Ferreira CNC-Center for Neuroscience and Cell Biology, University of Coimbra, Coimbra, Portugal

Eric Stice Oregon Research Institute, Eugene, OR, USA

Shahrad Taheri Department of Medicine, Weill Cornell Medical College—Qatar, Doha, Qatar

Anton B. Tonchev Laboratory of Cell Biology, Department of Anatomy and Histology, Medical University, Varna, Bulgaria

Giles S. H. Yeo MRC Metabolic Diseases Unit, Metabolic Research Laboratories, University of Cambridge, Cambridge, UK

Sonja Yokum Oregon Research Institute, Eugene, OR, USA

Chapter 1
Challenges in Understanding Development of Obesity

Thorkild I. A. Sørensen

The prevailing concepts about obesity is that it is a condition due to a positive whole body energy balance—determined by the difference between energy intake and energy expenditure—over a prolonged period of time. In this chapter, building on previously published reflections [1, 2], I will show how this intuitively obvious, but in reality naïve, concept can lead to fundamental flaws in our understanding of the development of obesity and hence potentially misguide the research in obesity as well as the management of the obesity problem in both preventive and therapeutic settings. The constructive outcome of this critique is an encouragement to shift focus to the regulation of the body fat mass.

The interest in understanding the development of obesity is of course primarily justified by the subsequent emergence of the health problems associated with obesity, in particular the metabolic abnormalities including ectopic fat deposition, insulin resistance, impaired glucose tolerance, dyslipidemia, and hypertension characterizing the metabolic syndrome as a major step toward the development of the various diseases associated with obesity, among many, diabetes and cardiovascular diseases in particular. These abnormalities may be induced by the limitations in adipose tissue expandability; when the process of development of obesity goes on exhausting the fat storage capacity, a chronic state of lipotoxicity and inflammatory response may ensue, which may induce the metabolic abnormalities constituting the metabolic syndrome. There may be important links between the processes of development of obesity and the development of the metabolic abnormalities, especially in relation to the determinants of the limits of the adipose tissue expandability. However, the processes behind development of the metabolic abnormalities will

---

T. I. A. Sørensen (✉)
Novo Nordisk Foundation Center for Basic Metabolic Research, Section on Metabolic Genetics, Faculty of Health and Medical Sciences, University of Copenhagen, Denmark, and Institute of Preventive Medicine, Bispebjerg and Frederiksberg Hospitals,
The Capital Region, Copenhagen, Denmark
e-mail: TSOE0005@regionh.dk

C. Nóbrega, R. Rodriguez-López (eds.), *Molecular Mechanisms Underpinning the Development of Obesity*, DOI 10.1007/978-3-319-12766-8_1

not be dealt with in this chapter, and interested readers are referred to previously published papers [3, 4].

The accumulation of fat as triglycerides in the adipose tissue is in principle a normal biological process, found ubiquitously in the animal world and serving several important needs in the body, of which the fat storage as reserve of energy for future use is an obvious one. At any given point in time it reflects the cumulative net balance between preceding normal, continuously operating fat dynamics. At the adipocyte level the dynamics of the balance is determined by uptake and synthesis of fatty acids (lipogenesis) and release of fatty acids (lipolysis). At the adipose tissue level, the dynamics is determined in addition by the current number of adipocytes, the development of new adipocytes from pre-adipocytes and stem cells, and apoptosis of adipocytes. On this background, it is a challenge to understand why the net balance of the fat dynamics may be altered both regarding the fat cell size determined by accumulated triglycerides and the adipocyte turnover at the tissue level. Very detailed and profound knowledge is available about the singular elements in these processes, both adipocyte function and adipocyte development, but the knowledge about the regulation of the integrated system, including the interplay with the rest of the body, is less well understood.

The fat dynamics is continuously fluctuating with various time constants (in relation to meals and circadian rhythm, by weekdays and weekends, by seasons, by developmental stages through the life cycle with obvious sex differences, by pregnancies, et cetera) but on average it keeps the balance with fairly stable fat mass over prolonged periods of time. However, on a daily basis the process of accumulation of body fat during development of obesity is so tiny—on average usually less than 1 % or even less than 0.5 % of total energy turnover—that it must be based on a very delicate imbalance in the integrated body systems that regulate the fat dynamics of the adipose tissue over longer periods of time. So, the crucial question is what is the difference in the long-term regulation of the fat dynamics between the normal situations where the balance is kept versus the situations where there is this tiny persisting imbalance leading to development of obesity? The accumulation of fat is so small that it cannot be measured as a positive energy balance by any of the available technics—only by a change in the amount of fat between two points in time, so far apart that the difference can be measured as a change in body composition. It is therefore a major challenge to study these differences when it can be measured only as differences in body composition over long time periods rather than as an ongoing energy imbalance.

The idea that obesity is 'due to' a positive whole body energy balance appears intuitively fairly obvious, but is in fact very unclear and possibly spurious in the way it is applied in the prevailing thinking about causes of obesity. A more precise and clear understanding of the relationship will undoubtedly help in understanding obesity development. According to the first thermodynamics law about conservation of energy, a body that has grown heavier because of accumulated body mass (apart from water accumulation) must have had a positive energy balance preceding the point in time it is measured as compared to its state before that growth in body mass. However, this mathematical-physical relationship between energy and body mass does not imply anything about the causal sequence or mechanisms that lead to the growth of the body mass. A causal inference is usually implicit in the words 'due to', and this confusion must be avoided. The relationship between the heavier body and the positive energy balance is purely descriptive and is indeed saying the same things with other words (a tautology). Thus, all combinations of processes that have induced alterations in the components of the energy balance equation on either side of the equation—energy storage equals energy intake minus energy expenditure—can in principle have been altered with the eventual outcome of a cumulative positive energy balance reflected in the accumulation of body mass. Thus, while the prevailing expectation is that the positive energy balance is due to an increased energy intake and/or a decreased energy expenditure, the primary causal mechanism could as well be the alteration of the regulation of the fat dynamics that subsequently leads to adjustments of the energy intake and expenditure quantitatively keeping the balance in accordance with the thermodynamic law.

The confusion about the relationship between obesity development and the positive energy balance is further exaggerated by the frequent mixing up the possible causes of the development with its effects on the energy balance. Obesity is usually associated with increases in fat-free components of the body that induce the increased energy needs. In most people developing obesity, the energy turnover increases because of increased energy intake serving the increased energy needs to sustain the obese body. Moreover, in accordance with the laws of physics any movement of the heavier body requires more energy, and on average the energy spent on physical activity usually is the same as for non-obese individuals. In contrast to the minute alterations in the energy balance components ending up in the accumulation of excess fat, these much larger effects that follow the development of obesity is clearly both measureable and often immediately visible. In the sense that the obese people are just satisfying their body needs for energy, they are not acting differently from non-obese people. It is a fundamental and regrettably frequent mistake to make the inference from the observation that the obese people eats more and move less, which is why they are obese. They do so to maintain the body functioning with an intact composition as if they were not obese. One could ask why the obese people are increasing the fat-free energy-demanding part of the body alongside with development of obesity and why they are not using the stored excess of energy in the adipose tissue to satisfy the increased needs of energy in the fat-free part of the body; there is no good answer to this question, and it is one of the fundamental challenges in understanding the development of obesity.

Another frequently encountered error in the arguments about the role of the energy balance and development of obesity is based on the fact that the whole body energy balance can be actively and voluntarily modified followed by a corresponding change in body mass and obviously in the amount of fat in the body. Thus, it is every one's experience from daily life that it is possible to actively enforce upon oneself a positive energy balance by overeating relative to the needs defined by the energy expenditure, or vice versa to actively enforce a negative energy balance leading to weight loss by restricting the food intake to an amount of energy that is less than the energy expenditure. In this situation, the adipose tissue acts as storage, the size of which passively reflects the enforced positive or negative energy balance. The argument then goes that since this is obviously possible for everyone to increase their body mass, and especially their body fat mass, by voluntary overeating, the development of obesity is likely based on the same mechanism of an enforced positive energy balance by overeating relative to the needs of energy. The claim implies that the obese people can be blamed for the development of their obesity because they have continuously been voluntarily overeating, and had they not chosen to do so, they would not have become obese. There are several reasons why this argument is false.

Experimental overfeeding studies showing the increase in body fat mass is based on an excess of energy intake that far exceeds the tiny amount of excess energy corresponding to the positive energy balance during obesity development. It makes no sense to blame those developing obesity for overeating an amount of energy that nobody can observe or even measure by the best available technics. Also these studies show that the body composition often is easily restored after cessation of the overfeeding, which is in stark contrast to the struggles people gaining weight on their way to the obese state have in their attempt to avoid the continued increase in body fat mass. Far the most obese people who do their best to get rid of the accumulated body fat by enforcing upon themselves a negative energy balance by eating less than the needs of energy (as determined by the increased energy turnover in their body) will sooner or later regain the body weight, restore the body fat mass and continue to grow if they were on that track before the intermittent departure. In other words, the development of obesity implies a drive to get this excess accumulation of fat alongside with the increase in the fat-free part of the body that is apparently similar to other basic inescapable physiologic drivers, for example breathing. It is a major challenge in obesity research to identify the biological mechanisms and determinants behind this drive toward an altered forcefully sustained body composition.

Insisting on the immediate link between a positive whole body energy balance and obesity development may also lead to another possibly fundamental error in the understanding of the process. The argument that there has been a cumulative energy balance preceding the state of an increased fat mass is true with regard to the fat mass as such, but not necessarily with regard to the whole body even though the fat-free part of the body usually also increases during obesity development. The disconnection of the whole-body energy balance from the body fat mass balance has been convincingly shown in experiments in obesity-prone mice [5]; a 5 % restriction in the daily energy supplies makes the fat depots grow in size at the expense of the non-fat part of the body (resulting in maintained body weight) compared with mice not exposed to this restriction. There may be various human analogies to this situation although none of them are so clear cut as the mouse experiments. American people living under food insecurity—defined as not knowing whether you have food for yourself and your family the next day—are the most obese. There is a great variation in the relation between the size of the fat and non-fat part of the human body, with some at one end of the range who have a clearly increased fat mass together with a small fat-free mass and even a lower than normal body weight. By age advancing above 65–70 years a process of increasing size of the fat mass concurrent with decreasing size of the fat-free part (sarcopenia) takes place. These scenarios show that it is necessary to keep a distinction between the energy balance of the fat mass and of the whole body, and it further emphasizes the need to approach the biological mechanisms regulating the size of the fat mass as possibly distinct from those regulating the size of other body components.

Although rather trivial, it should be mentioned that if people are living under such constraints in food supplies that neither the minute positive energy balance that is reflected in the increasing fat mass nor the concurrently developing increased fat-free mass and its energy requirement can be met, then the obesity development is blocked. The argument can then be switched around and it can be claimed that availability of food above this level is a prerequisite for obesity development, and, formally, availability of food can be defined as one necessary causal factor in obesity development. However, this self-evident argument does not help us to understand the individual differences in obesity development under the circumstances where there are no such constraints in food supplies. An important theoretical question is what happens to the body composition and function of those people who live under the constraints in food supplies that prevent obesity development, but who would have developed obesity had there not been such constraints. Do they as the obesity-prone mice develop increased size of their fat mass at the expense of the fat-free part of the body?

The expectation about a direct link between a positive whole-body energy balance and obesity development usually leads directly to focusing on the potentially modifiable behaviors that influence the whole-body energy balance, namely food intake and physical activity. When discussing why some people have developed obesity, the question emerges whether they have been eating too much or being too little physically active or both (often expressed in moral terms as gluttony, sloth or both). As it appears from the discussions above, this question is potentially misleading for several reasons. First, all observable differences between obese and non-obese people in this regard must take into account the effects of obesity on the behaviors. Second, if food intake and physical activity are considered as contributions to the energy balance purely by their energetic values, observation of an imbalance corresponding to the tiny positive energy balance reflecting the fat accumulation is completely out or reach by any measurement technics. Third, even if it had been technically possible to measure the energetic values of these behaviors with a precision allowing detection of the difference in energy balance during obesity development, it would not be possible to determine which changes might have induced the fat accumulation because of the concurrent changes in these same behaviors as consequence of the growth in the fat mass and fat-free mass and its implications for the energy intake and expenditure. In accordance with these conditions and relationships, it has been impossible to show that the food intake and physical activity levels in the non-obese state is associated with the future development of obesity.

However, the interest in food intake and physical activity may still be relevant in understanding obesity development, but then only by asking if the food intake, including its composition of foods and nutrients, and the physical activity, including for example distribution between sedentary periods and active periods with different types and intensity of physical activity, influence the fat dynamics by biological mechanisms beyond their energetic values? This is possible and plausible in view of the growing evidence indicating that excessive intake of sugar-sweetened beverages and prolonged completely sedentary life style may alter the fat dynamics toward an ongoing fat accumulation and eventual development of obesity.

The key challenge in understanding the development of obesity is not to make inferences from the simple understanding of the whole-body energy balance, but rather to unravel the changes in the biological mechanisms and the determinants of these changes leading to the persisting very small alterations in the fat dynamics resulting in an ongoing fat accumulation in the adipose tissue. While realizing this state-of-the-art, a very complex plethora of specific questions emerges; which biological mechanisms are altered, when during the life, where in the body, and how are the alterations induced and under which conditions. The following chapters will elucidate various aspects of this complexity.

While addressing these aspects it may be worthwhile to keep the perspectives in mind about the history of the development of the global obesity epidemic, defined as a steady increase in the occurrence of obesity in the world's populations. Obesity has been known as a body type for thousands of years in several cultures, but the epidemic appears to be a very recent phenomenon in populations that are not suffering from constraints in food supplies during the twentieth century, emerging in the United States of America in the first half of that century and in other parts of the world during the second half of the century. The epidemic has developed in a way that leaves no doubt that some fundamental changes in the broadly defined environment must have taken place that allows the obesity development. This so-called obesogenic environment is usually conceived as an environment where there is abundance of cheap, energy-dense foods and where it is easy to avoid or difficult to be physically active. However, in agreement with the discussion above, it remains an open question whether the transition to such environment plays a driving or just a permissive role in the global epidemic of obesity. As at the individual level, it has been difficult to show that the obesogenic changes of the environment preceded the takeoff in the epidemic. It is frequently claimed that the people who become obese in this environment are those genetically predisposed to obesity development, which indeed is a plausible explanation. On the other hand, some studies also suggest that the epidemic may be driven by changes in early life induced by more specific environmental influences, creating an additional predisposition either before conception, during pregnancy of during infancy in subgroups in the populations.

A major challenge in the understanding of development of obesity is to unite the evidence about the apparent multitude of determinants. There are sizeable differences in occurrence of obesity at different sites, even within small otherwise homogenous populations, and there are changes over time within the same populations that essentially remain unexplained. The last decade has shown a surprising heterogeneity, with a mixture of continued increases, leveling off, and even declines in the occurrence of obesity at different rates in different populations. Interrogation of these differences will likely pave the way to a much deeper understanding of development of obesity.

The key descriptive determinants of occurrence of obesity within given societies at given points in time are in addition to age and sex, also various psychosocial factors. Where there is abundance of foods available, the greatest risk of obesity development is among those who live with various psychosocial challenges. In contrast to the difficulties in showing relations of food intake and physical activity to later development of obesity, the prospective relation of psychosocial factors to obesity development is well confirmed. It has been tempting to claim that the psychosocial relations are attributable to obesogenic behaviors regarding food intake and physical activity, but this argument is invalid simply because it would require a stronger relation between the obesogenic behaviors and obesity development than with the psychosocial factors, which is not the case. There is no clear and convincing evidence explaining the role of the psychosocial factors, but it is tempting to hypothesize a direct link between the mental state induced by these conditions and—through a cerebral neuronal and/or hormonal connection—alterations in the fat dynamics in adipose tissue. The evolutionary justification of the process could be that since securing food supplies is dependent on collaborative activities in a well-functioning social environment, psychosocial challenges may be read by the mind as an indication of risks of deficient food supplies, and hence elicits the additional accumulation of fat as an energy reserve. Such processes could also be behind the early life influences as a mechanism securing energy also for the next generation. In the present context these speculations primarily demonstrates the necessity of taking a broader, cross-disciplinary approach, also when investigating the molecular mechanisms underpinning the development of obesity.

References

1. Sørensen TIA (2009) Challenges in the study of causation of obesity. Proc Nutr Soc 68:43–54
2. Sørensen TIA (2012) Is obesity a healthy active response to an expected future lack of energy rather than a passive storage of surplus energy? Obes Facts 5:431–435
3. Sørensen TIA, Virtue S, Vidal-Puig A (2010) Obesity as a clinical and public health problem: Is there a need for a new definition based on lipotoxicity effects? Biochim Biophys Acta 1801:400–404
4. Sørensen TIA (2011) Obesity defined as excess storage of inert triglycerides-so we need a paradigm shift? Obes Facts 4:91–94
5. Li X, Cope MB, Johnson MS, Smith DL Jr, Nagy TR (2010) Mild calorie restriction induces fat accumulation in female C57BL/6 J mice. Obesity (Silver Spring) 18:456−462

Chapter 2
Monogenic Forms of Obesity

Hélène Huvenne and Béatrice Dubern

Introduction

Obesity, defined as an excess fat mass, is a complex and multifactorial disease resulting from the interaction of numerous environmental and genetic factors. It is characterized by a wide phenotypic heterogeneity. Numerous epidemiological and intervention studies carried out in different cohorts (twins brought up together or separately, adopted children, nuclear families, among others) have recognized the role of individual genetic and biological susceptibilities in response to the current weight-gain promoting environment [1]. There are also individual differences in progression of weight (i.e. different trajectories) and risk of associated comorbidities. It is now well accepted that the development of obesity stems from the interaction of multiple environmental factors (such as overeating and/or reduction in physical activity) with genetic factors. The severity of obesity will be thus determined by the impact of environment on the genetic background of each individual.

B. Dubern (✉)
Nutrition and Gastroenterology Department, Armand-Trousseau Hospital, 26 Avenue du Dr Arnold Netter, 75012 Paris, France
e-mail: beatrice.dubern@trs.aphp.fr

Institute of Cardiometabolism and Nutrition, ICAN, Pitié-Salpêtrière Hospital, 47-83 Boulevard de l'hôpital, 75013 Paris, France

INSERM, UMR_S U1166, Nutriomics, Pitié-Salpêtrière Hospital, 47-83 Boulevard de l'hôpital, 75013 Paris, France

H. Huvenne
Pediatrics Department, Saint-Vincent de Paul Hospital, Boulevard de Belfort, 59000 Lille, France

Institute of Cardiometabolism and Nutrition, ICAN, Pitié-Salpêtrière Hospital, 47-83 Boulevard de l'hôpital, 75013 Paris, France

INSERM, UMR_S U1166, Nutriomics, Pitié-Salpêtrière Hospital, 47-83 Boulevard de l'hôpital, 75013 Paris, France

C. Nóbrega, R. Rodriguez-López (eds.), *Molecular Mechanisms Underpinning the Development of Obesity*, DOI 10.1007/978-3-319-12766-8_2

Several clinical presentations are described in obesity depending on the involved genes:

a. Monogenic obesity described as rare and severe early-onset obesity associated with endocrine disorders. There are mainly due to mutations in genes of the leptin/melanocortin axis involved in food intake regulation (genes of leptin (*LEP*) and its receptor (*LEPR*), proopiomelanocortin (*POMC*), proconvertase 1 (*PC1*), etc).
b. Syndromic obesity corresponding to obesity associated with others genetic syndromes. Patients are clinically severely obese and additionally distinguished by mental retardation, dysmorphic features and organ-specific developmental abnormalities. Prader-Willi and Bardet-Biedl syndromes are the 2 most frequent. But more than 100 syndromes are now associated to obesity.
c. Melanocortin 4 receptor (*MC4R*) linked obesity characterized by a variable severity of obesity and the absence of specific phenotype. They are responsible for 2–3 % of obesity in adults and children.
d. Polygenic obesity, which is the more common clinical situation (> 95 % of cases). Here each susceptibility gene, taken individually, would only have a slight effect on weight. The cumulative contribution of these genes would become significant only in an 'obesogenic lifestyle' (such as overfeeding, sedentariness, stress).

The comprehension of the obesity molecular mechanisms progressed enormously these last years thanks to the development of faster, precise and effective genetic screening tools. In particular, the whole-exome sequencing showed its power to identify new monogenic obesities due to mutations in the leptin/melanocortin pathway or in other genes. Rare genetic obesities are, in fact, important to clinically detect because it allows to progress in understanding obesity physiopathology and on the other hand there is a specific management of these forms of obesity depending on specific and multidisciplinary teams.

Monogenic Obesity Due to Mutations in the Leptin/Melanocortin Pathway

As described in rodents, monogenic obesities are mainly due to mutations in the genes encoding proteins involved in the leptin/melanocortin pathway that plays a pivotal role in the hypothalamic control of food intake (Fig. 2.1). The hypothalamic leptin/melanocortin pathway is activated following the systemic release of the adipokine LEP and its subsequent interaction with its receptor LEPR located on the surface of neurons of the arcuate nucleus. The downstream signals that regulate satiety and energy homeostasis are then propagated via POMC, cocaine-and-amphetamine-related transcript (CART) and the melanocortin system [2]. While POMC/CART neurons synthesize the anorectic peptide α-melanocyte stimulating hormone

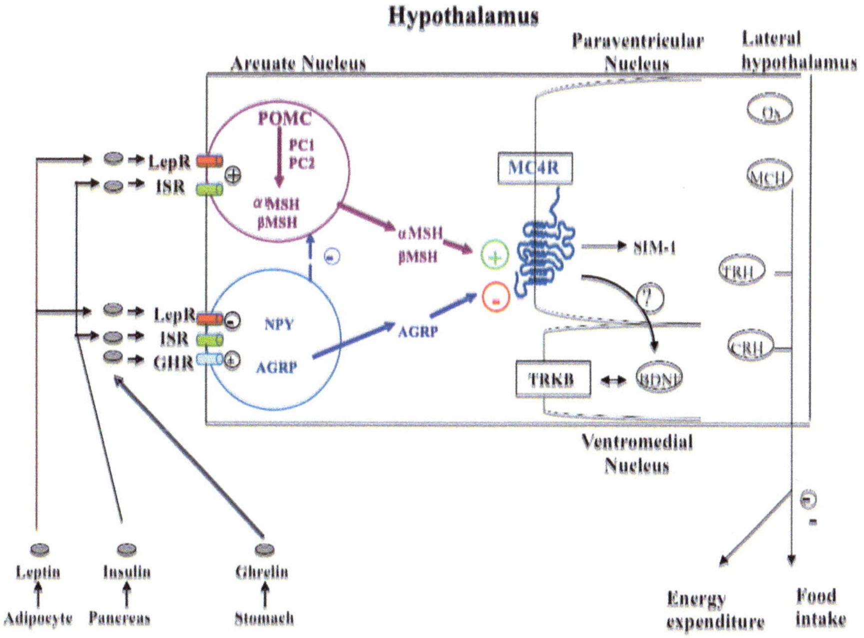

Fig. 2.1 The leptin/melanocortin pathway. Neuronal populations propagate the signaling of various molecules (leptin, insulin, ghrelin) to control food intake and satiety. POMC-neurons in the arcuate nucleus are activated by leptin and insulin and produce the α-MSH, which then activates the MC4R receptor in the paraventricular nucleus resulting in a satiety signal. The downstream roles of SIM1, BDNF and TKRB are currently being explored. A separate group of neurons expressing NPY and AGRP produce molecules that act as potent inhibitors of MC4R signaling. Several mutations of those genes involved in the leptin/melanocortin pathway are responsible for severe early-onset obesity. *AGRP* agouti-related protein, *BDNF* brain-derived neurotropic factor, *CRH* corticotrophin-releasing hormone, *GHR* ghrelin receptor, *ISR* insulin receptor, *LepR* leptin receptor, *MC4R* melanocortin-4 receptor, *MCH* melanin-concentrating hormone, *α-MSH* α-melanocyte stimulating hormone, *β-MSH* β -melanocyte stimulating hormone, *NPY* neuropeptide Y, *Ox* orexins, *PC1 and 2* proconvertase 1 and 2, *POMC* proopiomelanocortin, *SIM1* single-minded 1, *TRH* thyrotropin-releasing hormone, *TRKB* tyrosine kinase receptor

(α-MSH), a separate group of neurons express the orexigenic neuropeptide Y (NPY) and the agouti-related protein (AGRP), which acts as a potent inhibitor of melanocortin 3 (MC3R) and MC4R receptors. The nature of the POMC derived peptides depends on the type of endoproteolytic enzyme present in the specific brain region. In the anterior pituitary, the presence of the PC1 enzyme produces ACTH (adrenocorticotropic hormone) and β-lipotrophin peptides, while the combined presence of PC1 and PC2 in the hypothalamus controls the production of α-, β-, γ-MSH and β-endorphins. Mutations in human genes coding for LEP [3–5], LEPR [6, 7], POMC [8] and PC1 [9] lead to severe obesity occurring soon after birth, with generally complete penetrance and autosomal recessive transmission (Table 2.1).

Table 2.1 Rare monogenic forms of human obesity

Gene	Mutation type	Obesity	Associated phenotypes
Leptin (LEP)	Homozygous mutation	Severe, from the first days of life	Gonadotropic and thyrotropic insufficiency
Leptin receptor (LEPR)	Homozygous mutation	Severe, from the first days of life	Gonadotropic, thyrotropic and somatotropic insufficiency
Proopiomelanocortin (POMC)	Homozygous or compound heterozygous mutation	Severe, from the first months of life	ACTH insufficiency Mild hypothyroidism and ginger hairs if the mutation leads to the absence of POMC production
Proprotein convertase subtilisin/kexin type 1 (PCSK1)	Homozygous or compound heterozygous mutation	Severe obesity occurring in childhood	Adrenal, gonadotropic, somatotropic and thyrotropic insufficiency Postprandial hypoglycemic malaises Central diabetes insipidus
Single-minded 1 (SIM1)	Translocation between chr 1p22.1 and 6q16.2 in the SIM 1 gene	Severe obesity occurring in childhood	Inconstantly, neurobehavioral abnormalities (including emotional lability or autism-like behavior)
Neurotrophic tyrosine kinase receptor type 2 (NTRK2)	De novo heterozygous mutation	Severe obesity from the first months of life	Developmental delay Behavioral disturbance Blunted response to pain
Dedicator of cytokinesis 5 (DOCK5)	Variable number tandem repeats (VNTRs)	Childhood and adult severe obesity	–
Kinase suppressor of Ras2 (KSR2)	Heterozygous frameshift, nonsense or missense variants	Severe obesity	Hyperphagia in childhood Low heart rate Reduced basal metabolic rate Severe insulin resistance
Tubby-like protein (TUB)	Homozygous frameshift mutation	Early-onset obesity	Night blindness, decreased visual acuity and electrophysiological features of a rod cone dystrophy

Leptin Deficiency

Mutations in the human genes coding for LEP [3–5] and LEPR [6, 7] lead to rapid and dramatic increase in weight since the first months of life, as illustrated by the weight curve of LEPR deficient subjects (Fig. 2.2). Evaluating body composition in some *LEPR* mutation carriers show a large amount of total body fat mass (> 50 %) but resting energy expenditure remains related to the level of body corpulence. Feeding behaviour is characterized by major hyperphagia and ravenous hunger [10]. Surprisingly, one LEP deficient Austrian girl has been recently described with more moderate obesity (BMI 31.5 kg/m^2), despite an increased consumption

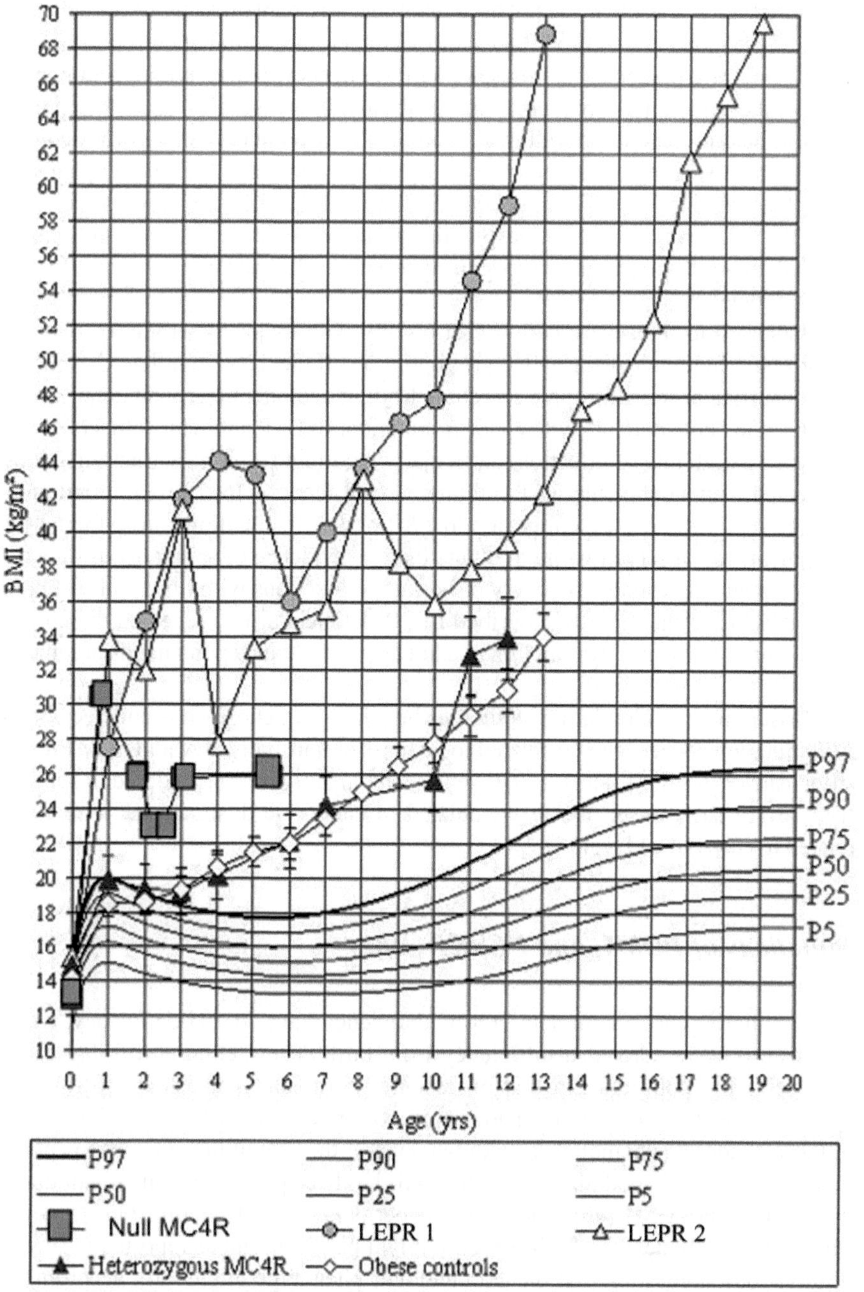

Fig. 2.2 BMI curves of 2 homozygous null *LEPR* mutants (LEPR 1 and 2), 1 homozygous null *MC4R* patient, 6 heterozygous *MC4R* carriers and 40 non mutated obese controls [47]. The reference curves are the standard French/Institut National de la Santé et de la Recherche Médicale percentile curves

of calories in a test meal [11]. The phenotype was explained by extremely low daily calorie intake. Even if one takes into account a substantial underreporting, this observation might suggest that despite LEP deficiency, it was possible to control energy intake and thus to prevent extreme obesity. In that specific case, the parents' role was determinant by providing a favourable environment with vigorous control of the patient's eating behaviour from early infancy onward. A further explanation might be related to the different genetic backgrounds of different subjects with LEP or LEPR deficiency. However, despite this particular case, severe early-onset obesity with major hyperphagia is recognized as a main clinical presentation of LEP or LEPR deficiency.

Associated to the severe early-onset obesity with major hyperphagia, hypogonadotrophic hypogonadism and thyrotrophic insufficiencies complete the phenotype. Insufficient somatotrophic secretion, leading to moderate growth delay, is also described in some patients with a *LEPR* mutation. In LEP deficient subjects, it was described a high rate of infection, particularly recurrent respiratory tract infections, associated with a deficiency in T cell number and function [6, 12, 13]. In individuals with leptin deficiency either due to *LEP* or *LEPR* mutations, no pubertal development was observed while in others there is evidence of spontaneous pubertal development suggesting a recovery of hormonal functions with time. For example, the follow-up of the initially described LEPR deficient sisters revealed the normalization of thyroid mild dysfunction at adult age and normal spontaneous pregnancy [14].

Measurement of circulating leptin may help in the diagnosis: it is undetectable in *LEP* mutation carriers and correlated to fat mass or unusually elevated in *LEPR* mutation carriers [3, 6, 7]. Thus, *LEPR* gene screening might be considered in subjects with severe early-onset obesity associated to endocrine dysfunctions with leptin related to corpulence level [7].

Mutations of POMC and PCSK1 Genes

Obese children with a complete POMC deficiency have ACTH deficiency, which can lead to acute adrenal insufficiency from birth. These children display also a mild central hypothyroidism that necessitates hormonal replacement [8]. Alterations in the somatotropic and gonadotropic axis are also described [15]. The reason of these endocrine anomalies is unknown even if the role of melanocortin peptides in influencing the hypothalamic pituitary axis has been proposed. Ginger hair due to the absence of α-MSH, which activates the peripheral MC1R involved in pigmentation is classically described. However, it might be inconstant as reported in several observations suggesting that the skin and hair phenotype might vary according to the ethnic origin of *POMC* mutation carriers [15–17]. The modifications in color hair, adrenal function and body weight are consistent with the lack of POMC-derived ligands for the melanocortin receptors MC1R, MC2R and MC4R respectively.

Patients carrying rare mutation in the *PCSK1* (proprotein convertase subtilisin/kexin type 1) gene leading to PC1 deficiency, have also endocrine anomalies in addition to severe obesity. They are mainly postprandial hypoglycemic malaises, fertility disorders due to hypogonadotrophic hypogonadism, central hypothyroidism and adrenal insufficiency secondary to lack of POMC maturation. The delayed postprandial malaises are explained by the accumulation of proinsulin through lack of PC1, which is involved in the synthesis of mature insulin from proinsulin. The absence of POMC maturation causes a dysfunction in the melanocortin pathway that explains the obese phenotype [9]. Severe persistent diarrhea, due to defect in intestinal absorption, is also described, secondary to lack in mature GLP-1 (glucagon-like peptide-1) [18]. Alteration of the processing of prohormones, progastrin and proglucagon, explains, at least in part, the intestinal phenotype and also suggests the role of PC1 in absorptive functions in the intestine. Recently, central diabetes insipidus improved by oral desmopressin was noted in one compound heterozygous proband and in 13 children with a total PC1 deficiency. These observations suggest that PCSK1 may be involved in the full functioning or central sensing of osmolality in humans [19, 20]. Growth hormone deficiency was also noted in the 13 children with complete PC1 deficiency [20].

Diagnosis Genetic Testing of Monogenic Obesity

In case of clinical situation suggesting a monogenic obesity (severe early-onset obesity associated to endocrine anomalies and potentially consanguineous parents), direct sequencing of the candidate gene (*LEP*, *LEPR*, *POMC*, etc.) is necessary for diagnosis confirmation. It will detect homozygous or compound heterozygous mutation responsible for an interruption of the leptin-melanocortin axis. Family members are needed to be tested for segregation analysis and to evaluate the risk of recurrence.

A few genetics laboratories routinely perform those analyses that usually are part of research programs. For example in Europe, this genetic testing cans be found at

- UF nutrigénétique (Pitié-Salpêtrière hospital). Contact: B. Dubern or K. Clément. Address: Endocrinology and Nutrition department, Pitié-Salpêtrière hospital, Boulevard de l'hôpital, 75013 Paris or Pediatric Nutrition department, Trousseau hospital, 75012 Paris. Email: beatrice.dubern@trs.aphp.fr; karine.clement@psl.aphp.fr; Tel: 33 (0) 14234 8936; Fax: 33 (0) 4051 00 57; Web site: http//www.cgmc-psl.fr/
- S O'Rahilly's team (University of Cambridge). Contact: S. Farooqi or S. O'Rahilly. Address: Department of Clinical Biochemistry, University of Cambridge, Addenbrooke's Hospital, Hills Road, Cambridge, CB2 2QR. Email: isf20@cam.ac.uk; Tel: +44 (0) 1223 336792; Fax: +44 (0) 1223 330598; Web site: http://www.mrl.ims.cam.ac.uk

Treatment

In case of LEP deficiency, children and adults benefit from subcutaneous injection of leptin, resulting in weight loss, mainly of fat mass, with a major effect on reducing food intake and on other dysfunctions including immunity, as described previously [21]. A detailed microanalysis of eating behavior of three leptin deficient adults, before and after leptin treatment, revealed reduced overall food consumption, a slower rate of eating and diminished duration of eating of every meal in the three subjects after leptin therapy. Leptin treatment also induces features of puberty even in adults [10]. This study supports a role of leptin in influencing the motivation to eat before each meal [22]. Another study shows that leptin treatment had a major effect on food intake. *Ad libitum* energy intake in a test meal was reduced, hunger ratings in the fasted state decreased and satiety following a meal increased. Leptin acts on neural circuits governing food intake to diminish perception of food reward while enhancing the response to satiety signals generated during food consumption [23]. In a separate study, hormonal and metabolic changes were evaluated before and after leptin treatment [10]. Leptin treatment was able to induce aspects of puberty even in adults, as illustrated by the effect of leptin treatment in one 27 year-old adult male with hypogonadism. In two women between 35–40 years, leptin treatment led to regular menstrual periods and hormonal peaks of progesterone evoking a pattern of ovulation. Although cortisol deficiency was not initially found in LEP deficient patients, eight months of leptin treatment modified the pulsatility of cortisol with a greater morning rise of cortisol. Leptin could have a previously unsuspected impact on human hypothalamic-pituitary-adrenal function. Metabolic parameters of leptin deficient patients improved in parallel with weight loss.

In the LEPR deficient subjects, leptin treatment is useless because of a non-functional LEPR. Factors that could possibly bypass normal leptin delivery systems are being developed but are not yet currently available. The ciliary neurotrophic factor (CNTF) was one of the candidate molecules. CNTF activates downstream signaling molecules such as STAT-3 in the hypothalamus area that regulates food intake, even when administered systemically. Treatment with CNTF in humans and animals induced substantial loss of fat mass [24]. The neurotrophic factor, Axokine, an agonist for the CNTF receptor, has been under development by the Regeneron Company, for the potential treatment of obesity and its metabolic associated complications. But the phase III clinical trials were stopped due to development of antibodies against Axokine in nearly 70 % of the tested subjects after approximately 3 months of treatment. In addition, Axokine had a small positive effect [25]. It is also possible that side effects of CNTF, a molecule possibly acting in the immune function, might be expected [26].

In children with a complete POMC deficiency, a 3 months trial using MC4R agonist with low affinity was inefficient on weight or food intake [8]. POMC, PC1 or LEPR deficient families might benefit from the development of new MC4R agonists if such drugs become available, in order to restore the melanocortin signal. Likewise, deep brain stimulation trials with an electrode inserted in the anterior third ventricle contiguous to the ventromedian hypothalamus are actually performed. In monkeys, this chronic 8-week stimulation induces a significant decrease in corpulence with reduction of 8% of body weight and 18% of fat mass, without side effects [27].

Today, bariatric surgery is the only long-term efficient treatment for severe obesity [28] using several surgical technics (laparoscopic gastric bypass, gastric banding or sleeve gastrectomy). The question of such treatment and its potential efficiency is crucial in patients with monogenic obesity. But currently, data on bariatric surgery in these patients are limited and controversial. In one LEPR deficient patient, vertical gastroplasty was beneficial and sufficient to induce and maintain a 40 kg weight loss (−20% of the initial weight) over 8 years of regular follow-up, whereas the patient remained obese [29]. In contrast, a relative failure of surgical therapy was illustrated by the rapid weight regain 1 year after bypass in another LEPR deficient morbidly obese woman. But this patient with low socioeconomic status had extreme difficulties after postsurgical counselling. She was noncompliant with the recommendations provided in this type of purely restrictive surgery and her medical follow-up was very irregular. This report illustrated the important role of environment on the benefice of bariatric surgery especially in case of monogenic obesities or underlined the poor efficiency of bariatric surgery in these patients [29].

So, due to the limited number of cases, the long-term efficacy and safety of bariatric surgery need further evaluation. A multidisciplinary team approach should always be adopted in order to establish the correct indication and realistic explanation after surgical treatment of obese patients.

Other Monogenic Obesities

Several additional genes, implicated in the hypothalamus and central nervous system development, have been found to cause monogenic obesity in humans. A deletion of the *SIM1* (single-minded homolog 1) gene, located on the 6q chromosome, secondary to a *de novo* translocation between 1p22.1 and 6q16.2 chromosomes, was identified in one girl with severe early-onset obesity associated to hyperphagia and food impulsivity [30]. She had a rate of early weight gain comparable to the weight curve of LEP and LEPR deficient children. Izumi et al. identified also an interstitial 6q deletion including the *SIM1* gene in a subject with Prader-Willi-like features (neonatal hypotonia, dysmorphy, developmental delay, early-onset obesity, short stature, hypopituitarism) [31]. *SIM1* encodes a transcriptional factor implicated, in mouse, in the development of the hypothalamic paraventricular nucleus. It plays a role in the melanocortin signaling pathway and appears to regulate feeding rather than energy expenditure [32, 33]. The sequencing of the coding region of *SIM1*, in 2100 unrelated patients with severe early-onset obesity and in 1680 unrelated population-based controls, identified 13 different heterozygous variants in 28 severely obese patients. Variants carriers exhibited severe obesity, increased *ad libitum* food intake in a test meal, normal basal metabolic rate and inconstantly neurobehavioral phenotype (impaired concentration, memory deficit, emotional liability or autistic spectrum behavior). Nine of the 13 variants significantly reduced the ability of *SIM1* to activate a *SIM1*-responsive reporter gene. These mutations co-segregated with obesity in extended family studies with variable penetrance. So, rare variants in *SIM1* should be considered in patients with hyperphagic early-onset obesity associated or not to Prader-Willi-like syndrome features or neurobehavioral abnormalities such as emotional lability or autism-like behavior [34–36].

Likewise, decreased expression of the brain-derived neurotropic factor (BDNF) was found to regulate eating behavior [37]. BDNF, encoded by *NTRK2* (neurotrophic tyrosine kinase receptor type 2) gene, and its associated tyrosine kinase receptor (TRKB) are both expressed in the ventromedial hypothalamus. They have been attributed a role downstream of MC4R signaling implicated in feeding regulation [38]. A *de novo* heterozygous mutation in *NTRK2* gene was described in an 8-year-old boy with severe early-onset obesity, mental retardation, developmental delay and anomalies of higher neurological functions, like the impairment of early memory, learning and nociception [39]. Other mutations in *NTRK2* were found in patients with early-onset obesity and developmental delay, but their functional consequences and their implication in obesity are yet to demonstrate. In vitro studies of some mutations have suggested that these mutations could impair hypothalamic-signaling processes [40].

Finally, the contribution of copy number variants (CNVs) to complex disease susceptibility, such as severe obesity, has been the subject of debates in recent years. Variable number tandem repeats (VNTRs) constitute a relatively under-examined class of genomic variants in the context of complex disease. Rare CNVs have been shown to be responsible for severe highly penetrant forms of obesity. For example, investigation of a complex region on chromosome 8p21.2 encompassing the *DOCK5* (dedicator of cytokinesis 5) gene has shown a significant association of the *DOCK5* VNTRs with childhood and adult severe obesity [41]. So, more systematic investigation of the role of VNTRs in obesity had to be performed to study their relatively unexplored contribution to this disease and their potential link with the leptin/melanocortin pathway.

The rapid development of new tools such as whole-exome sequencing will probably help to identify novel genes in severe early-onset obesity or monogenic obesity. The whole-exome sequencing is a diagnostic approach for identification of molecular defects in patients with suspected genetic disorders. It showed its power to identify mutations responsible for rare diseases, in a small number of unrelated affected individuals. Indeed, it contributed greatly to the discovery of disease-causing genes in several rare inherited human diseases. So, this tool could probably help to reveal new molecular abnormalities in patients with monogenic obesity. It was tested and validated in a study for the molecular diagnosis of 43 forms of monogenic diabetes or obesity. Forty patients (19 with a monogenic form of diabetes and 21 with a monogenic form of obesity) carrying a known causal mutation for those subtypes according to diagnostic laboratories were blindly re-analyzed. Except for one variant, all causal mutations in each patient were re-identified, associated with an almost perfect sequencing of the targets (mean of 98.6 %). Moreover, in three individuals, other mutations were detected with a putatively deleterious effect, in addition to those previously reported by the genetic diagnosis laboratories [42]. In another study based on 39 unrelated severely obese Pakistani children, the whole-exome sequencing revealed two novel homozygous *LEPR* mutations in two probands who were phenotypically indistinguishable from age-matched LEP deficient subjects of the same population [43]. Several other genes implicated in severe obesity have been identified using this method: one paternal 2pter deletion, encompassing the *ACP1* (acid phosphatase 1), *TMEM18* (transmembrane protein 18) and *MYT1L*(myelin transcription factor 1-like) genes in five unrelated patients presenting with severe early-onset obesity, hyperphagia, intellectual deficiency and severe behavioral difficulties [44]; multiple rare variants in the *KSR2* (kinase supressor of Ras 2) gene in 45 unrelated severely obese individuals exhibiting low heart rate, reduced basal metabolic rate and severe insulin resistance [45]; one homozygous frameshift mutation in the *TUB* (tubby-like protein) gene in a proband who presented with obesity, night blindness, decreased visual acuity, and electrophysiological features of rod cone dystrophy [46].

In conclusion, whole-exome sequencing will probably help physicians to identify new molecular abnormalities in patients with severe early-onset obesity in a close future. Moreover, progress in understanding the monogenic obesity mechanisms may help to better understand the pathophysiology of the more common forms of obesity and to improve their management.

References

1. Farooqi IS (2006) Genetic aspects of severe childhood obesity. Pediatr Endocrinol Rev 3(Suppl 4):528–536
2. Harrold JA, Williams G (2006) Melanocortin-4 receptors, beta-MSH and leptin: key elements in the satiety pathway. Peptides 27:365–371
3. Montague CT, Farooqi IS, Whitehead JP, Soos MA, Rau H, Wareham NJ et al (1997) Congenital leptin deficiency is associated with severe early-onset obesity in humans. Nature 387:903–908
4. Strosberg AD, Issad T (1999) The involvement of leptin in humans revealed by mutations in leptin and leptin receptor genes. Trends Pharmacol Sci 20:227–230
5. Fatima W, Shahid A, Imran M, Manzoor J, Hasnain S, Rana S et al (2011) Leptin deficiency and leptin gene mutations in obese children from Pakistan. Int J Pediatr Obes 6:419–427
6. Clément K, Vaisse C, Lahlou N, Cabrol S, Pelloux V, Cassuto D et al (1998) A mutation in the human leptin receptor gene causes obesity and pituitary dysfunction. Nature 392:398–401
7. Farooqi IS, Wangensteen T, Collins S, Kimber W, Matarese G, Keogh JM et al (2007) Clinical and molecular genetic spectrum of congenital deficiency of the leptin receptor. N Engl J Med 356:237–247
8. Krude H, Biebermann H, Schnabel D, Tansek MZ, Theunissen P, Mullis PE et al (2003) Obesity due to proopiomelanocortin deficiency: three new cases and treatment trials with thyroid hormone and ACTH4-10. J Clin Endocrinol Metab 88:4633–4640
9. Jackson RS, Creemers JW, Ohagi S, Raffin-Sanson ML, Sanders L, Montague CT et al (1997) Obesity and impaired prohormone processing associated with mutations in the human prohormone convertase 1 gene. Nat Genet 16:303–306

10. Licinio J, Caglayan S, Ozata M, Yildiz BO, de Miranda PB, O'Kirwan F et al (2004) Phenotypic effects of leptin replacement on morbid obesity, diabetes mellitus, hypogonadism, and behavior in leptin-deficient adults. Proc Natl Acad Sci U S A 101:4531–4536
11. Fischer-Posovszky P, von Schnurbein J, Moepps B, Lahr G, Strauss G, Barth TF et al (2010) A new missense mutation in the leptin gene causes mild obesity and hypogonadism without affecting T cell responsiveness. J Clin Endocrinol Metab 95:2836–2840
12. Farooqi IS (2005) Genetic and hereditary aspects of childhood obesity. Best Pract Res Clin Endocrinol Metab 19:359–374
13. Mazen I, El-Gammal M, Abdel-Hamid M, Farooqi IS, Amr K (2011) Homozygosity for a novel missense mutation in the leptin receptor gene (P316T) in two Egyptian cousins with severe early onset obesity. Mol Genet Metab 102:461–464
14. Nizard J, Dommergues M, Clément K (2012) Pregnancy in a woman with a leptin-receptor mutation. N Engl J Med 366:1064–1065
15. Clément K, Dubern B, Mencarelli M, Czernichow P, Ito S, Wakamatsu K et al (2008) Unexpected endocrine features and normal pigmentation in a young adult patient carrying a novel homozygous mutation in the POMC gene. J Clin Endocrinol Metab 93:4955–4962
16. Farooqi IS, Drop S, Clements A, Keogh JM, Biernacka J, Lowenbein S et al (2006) Heterozygosity for a POMC-null mutation and increased obesity risk in humans. Diabetes 55:2549–2553
17. Carroll L, Voisey J, van Daal A (2005) Gene polymorphisms and their effects in the melanocortin system. Peptides 26:1871–1885
18. Jackson RS, Creemers JW, Farooqi IS, Raffin-Sanson ML, Varro A, Dockray GJ et al (2003) Small-intestinal dysfunction accompanies the complex endocrinopathy of human proprotein convertase 1 deficiency. J Clin Invest 112:1550–1560
19. Frank GR, Fox J, Candela N, Jovanovic Z, Bochukova E, Levine J et al (2013) Severe obesity and diabetes insipidus in a patient with PCSK1 deficiency. Mol Genet Metab 110:191–194
20. Martin MG, Lindberg I, Solorzano-Vargas RS, Wang J, Avitzur Y, Bandsma R et al (2013) Congenital proprotein convertase 1/3 deficiency causes malabsorptive diarrhea and other endocrinopathies in a pediatric cohort. Gastroenterology 145:138–148
21. Farooqi IS, Matarese G, Lord GM, Keogh JM, Lawrence E, Agwu C et al (2002) Beneficial effects of leptin on obesity, T cell hyporesponsiveness, and neuroendocrine/metabolic dysfunction of human congenital leptin deficiency. J Clin Invest 110:1093–1103
22. Williamson DA, Ravussin E, Wong ML, Wagner A, Dipaoli A, Caglayan S et al (2005) Microanalysis of eating behavior of three leptin deficient adults treated with leptin therapy. Appetite 45:75–80
23. Farooqi IS, Bullmore E, Keogh J, Gillard J, O'Rahilly S, Fletcher PC (2007) Leptin regulates striatal regions and human eating behavior. Science 317:1355
24. Sleeman MW, Anderson KD, Lambert PD, Yancopoulos GD, Wiegand SJ (2000) The ciliary neurotrophic factor and its receptor, CNTFR alpha. Pharm Acta Helv 74:265–272
25. Preti A (2003) Axokine (Regeneron). IDrugs 6:696–701
26. Sariola H (2001) The neurotrophic factors in non-neuronal tissues. Cell Mol Life Sci 58:1061–1066
27. Torres N, Chabardes S, Piallat B, Devergnas A, Benabid AL (2012) Body fat and body weight reduction following hypothalamic deep brain stimulation in monkeys: an intraventricular approach. Int J Obes 36:1537–1544
28. Sjöström L (2013) Review of the key results from the Swedish Obes Subjects (SOS) trial—a prospective controlled intervention study of bariatric surgery. J Intern Med 273:219–234
29. Le Beyec J, Cugnet-Anceau C, Pépin D, Alili R, Cotillard A, Lacorte JM et al (2013) Homozygous leptin receptor mutation due to uniparental disomy of chromosome 1: response to bariatric surgery. J Clin Endocrinol Metab 98:397–402
30. Holder JL Jr Butte NF Zinn AR (2000) Profound obesity associated with a balanced translocation that disrupts the SIM1 gene. Hum Mol Genet 9:101–108

31. Izumi K, Housam R, Kapadia C, Stallings VA, Medne L, Shaikh TH et al (2013) Endocrine phenotype of 6q16.1-q21 deletion involving *SIM1* and Prader-Willi syndrome-like features. Am J Med Genet 9999:1–7
32. Michaud JL, Boucher F, Melnyk A, Gauthier F, Goshu E, Levy E et al (2001) Sim1 haploinsufficiency causes hyperphagia, obesity and reduction of the paraventricular nucleus of the hypothalamus. Hum Mol Genet 10:1465–1473
33. Kublaoui BM, Holder JL Jr, Gemelli T, Zinn AR (2006) Sim1 haploinsufficiency impairs melanocortin-mediated anorexia and activation of paraventricular nucleus neurons. Mol Endocrinol 20:2483–2492
34. Ramachandrappa S, Raimondo A, Cali AMG, Keogh JM, Henning E, Saeed S et al (2013) Rare variants in single-minded 1 (SIM1) are associated with severe obesity. J Clin Invest 123:3042–3050
35. Bonnefond A, Raimondo A, Stutzmann F, Ghoussaini M, Ramachandrappa S, Bersten DC et al (2013) Loss-of-function mutations in *SIM1* contribute to obesity and Prader-Willi-like features. J Clin Invest 123:3037–3041
36. Zegers D, Beckers S, Hendrickx R, Van Camp JK, de Craemer V, Verrijken A et al (2014) Mutation screen of the SIM1 gene in pediatric patients with early-onset obesity. Int J Obes 38:1000–1004
37. Kernie SG, Liebl DJ, Parada LF (2000) BDNF regulates eating behavior and locomotor activity in mice. Embo J 19:1290–1300
38. Xu B, Goulding EH, Zang K, Cepoi D, Cone RD, Jones KR et al (2003) Brain-derived neurotrophic factor regulates energy balance downstream of melanocortin-4 receptor. Nat Neurosci 6:736–742
39. Yeo GS, Connie Hung CC, Rochford J, Keogh J, Gray J, Sivaramakrishnan S et al (2004) A de novo mutation affecting human TrkB associated with severe obesity and developmental delay. Nat Neurosci 7:1187–1189
40. Gray J, Yeo G, Hung C, Keogh J, Clayton P, Banerjee K et al (2007) Functional characterization of human NTRK2 mutations identified in patients with severe early-onset obesity. Int J Obes (Lond) 31:359–364
41. El-Sayed Moustapha JS, Eleftherohorinou H, de Smith AJ, Andersson-Assarsson JC, Couto Alves A, Hadjigeorgiou E et al (2012) Novel association approach for variable number tandem repeats (VNTRs) identifies *DOCK5* as a susceptibility gene for severe obesity. Hum Mol Genet 21:3727–3738
42. Bonnefond A, Philippe J, Durand E, Muller J, Saeed S, Arslan M et al (2014) Highly sensitive diagnosis of 43 monogenic forms of diabetes or obesity, through one step PCR-based enrichment in combination with next-generation sequencing. Diabetes Care 37:460–467
43. Saeed S, Bonnefond A, Manzoor J, Philippe J, Durand E, Arshad M et al (2013) Novel LEPR mutations in obese Pakistani children identified by PCR-based enrichment and next generation sequencing. Obesity 22(4):1112–1117
44. Doco-Fenzi M, Leroy C, Schneider A, Petit F, Delrue MA, Andrieux J et al (2013) Early-onset obesity and paternal 2pter deletion encompassing the *ACP1*, *TMEM18*, and *MYT1L* genes. Eur J Hum Genet 22:1–9
45. Pearce LR, Atanassova N, Banton MC, Bottomley B, van der Klaauw AA, Revelli JP et al (2013) KSR2 mutations are associated with obesity, insulin resistance, and impaired cellular fuel oxidation. Cell 155:765–777
46. Borman AD, Pearce LR, Mackay DS, Nagel-Wolfrum K, Davidson AE, Henderson R et al (2014) A homozygous mutation in the *TUB* gene associated with retinal dystrophy and obesity. Hum Mutat 35:289–293
47. Lubrano-Berthelier C, Dubern B, Lacorte JM, Picard F, Shapiro A, Zhang S et al (2006) Melanocortin 4 receptor mutations in a large cohort of severely obese adults: prevalence, functional classification, genotype-phenotype relationship, and lack of association with binge eating. J Clin Endocrinol Metab 91:1811–1818

Chapter 3
Heterogeneous Obesity Syndromes: New Strategies for Diagnosis

Raquel Rodríguez-López

Array Techniques for Diagnosis of New Obesity Syndromes

Obesity is a multifactorial and heterogeneous condition whose inheritance patterns are results from alterations of various genes, each having a complete, partial and/or additive effect. The identification of susceptibility genes and genetic variants for obesity requires various methodological approaches and to date have been classified into three main categories on the basis of genetic etiology: polygenic obesity (in which the combination of low number of single nucleotide polymorphisms (SNPs) have already been considered the molecular basis of risk genetic score), monogenic non syndromic (around 4–5 % Mendelian inherited cases due to high penetrance mutations), and syndromic obesity.

Considering an individual as candidate or not to study for genetic susceptibility to obesity, percentile or Body mass index (BMI) for age and percentile may be calculated for child or individuals more than 16 years of age, respectively. The consensus protocol to diagnose patients who have developmental delays or intellectual disabilities and other variable phenotypic features, requires a detailed clinical examination and precise description of the phenotype and a wide pedigree with an exhaustive familial antecedent. The screening may be standardized with radiological (hands, feet, spine, and renal ultrasonography) and biological investigation (hormonal, metabolic). After clinical evaluation, directed genetic analysis looking for suspected syndromic obesogenic etiologies as Prader Willi, Bardet-Biedl among others may be previously discarded. After discarding the clinically recognizable main monogenic etiologies of syndromic obesity, the phenotypic description of "new" syndromes with obesity related to chromosomal anomalies should point out new candidate genes implicated in the development of severe overweight.

R. Rodríguez-López (✉)
Laboratory of Molecular Medicine, Clinical Analysis Service,
Consorcio Hospital General de Valencia, Valencia, Spain
e-mail: rodriguez_raqlop@gva.es

C. Nóbrega, R. Rodriguez-López (eds.), *Molecular Mechanisms Underpinning the Development of Obesity*, DOI 10.1007/978-3-319-12766-8_3

Since the end of 2011, whole genome analysis using SNPs/CNVs and Comparative Genomic Hybridization array (CGH) analysis constitutes the first line testing for the most unaffiliated developmental delay syndromes. These techniques are already established into the clinical setting as the principal tool to detect human variation in routinely clinical assistance [1]. The array technique analysis has detected the presence of highly variable number of CNVs per subject. The International Standards for Cytogenomic Arrays (ISCA) Consortium promotes standardization of variant interpretation using standard filters, which have been established to consider them as pathogenic or not: (a) the rate of overlap with known nonpathogenic polymorphic CNVs, (b) smaller than 100 kb, (c) if the CNV did not involve known genes, (d) detected with a low probe density <2 kb, except for CNVs affecting regions known to be involved in microdeletion or microduplication. The test may be expanded to parents to distinguish between de novo-generated and inherited deleterious imbalances. The information is been continuously included on the similarly named and publicly available database (http://www.iscaconsortium.org).

Committees of clinical genetics laboratory experts have created and maintain publicly available databases at NCBI/NIH (http://www.ncbi.nlm.nih.gov), including the cytogenetic array data generated in clinical testing laboratories as the ISCA data and others as Unique (http://www.rarechromo.org), The database of Genomic Variants (http://projects.tcag.ca), UCSC Genome browser (http://www.genome.ucsc.edu) and DECIPHER (http://www.sanger.ac.uk/PostGenomics/decipher). All of them constitute the standards and the main recommendations for the interpretation and reporting of pathogenic versus benign copy number changes, as well as imbalances of unknown clinical significance in the context of the clinical assistance routine. The effort of clinicians and researchers to share their knowledge of copy number variation (CNV) among individuals with abnormal features, conform an invaluable resource in a process to establish sure guidelines to improve patient diagnosis. An adequate validation of the generated information will depend on the availability, accuracy and accessibility on genotype-phenotype relationships information. Interactive databases on Internet offer a suite of tools designed to facilitate the clinical interpretation of chromosomal imbalance.

Considering an obese patient as candidate to be studied using array techniques, may be established as obesity above 90th percentile under 14 year-old and at least one criterion among mental retardation and/or facial dimorphism [1]. Also, one major malformation may exist as uro-genital, cardiac, skeletal, cerebral or ophthalmologic, among others. Familial aggregation pattern of common obesity and/or obesity antecedents with an identified etiology, have also to be controlled. Nowadays, the utility of genetic testing using the array technologies on individuals with those phenotypes, even including autism, is absolutely accepted.

Multidisciplinary groups of experts in clinical genetics assistance, cytogenetic and molecular techniques, genomics and bioinformatics, participate in a dynamic process to design increasingly higher density arrays for general appliance, but also directed to improve the efficiency to detect the causes of specific phenotypes. This strategy is based on reviewing the data from chromosome array testing data of individuals with the same signs and symptoms. Experts designed common formularies to fill the same set of phenotype features and characteristics of each analyzed patient, however there are no options to reflect specific parameters to define obesity in those questionnaires. In addition to this, the array techniques recommendations have been mainly focused on clinical indications for pediatric applications. Babies or children with unexplained developmental disabilities, exhibit the majority of signs and/or symptoms of their global features, which compose their phenotypes at birth time, preferentially. However, obesity needs a variable but always a longer period time to appear.

Instead of the deficient and unequal information related to overweight in that series of patients, different syndromes with obesity have been associated with chromosomal imbalance found by arrays. Their pathogenicity for obesity susceptibility has not been considered definitely proved in all patients; further specifically designed studies are required to corroborate the genotype-phenotype correlations for specific involved genes and mechanisms responsible for the obesity. Wide studies in cohorts of patients selected for syndromic obesity would allow the characterization of cryptic chromosomal anomalies and identify candidate genes present at the deleted or duplicated *loci*.

Chromosomal Imbalance Affecting High Susceptibility Genes to Obesity

Generally speaking about risk genes to obesity, the previous genetic analysis of some chromosomal regions harboring one or more genes among members belonging to families with multiple cases, and test whether co-segregate with the trait, identified few variants with high penetrance [2, 3]. Those results do not contribute substantially to the risk of common forms of obesity [4, 5] but constitute the sufficient molecular cause of the obesity appearance among individuals who carry genetic alterations on these genes. Haploinsufficiency of these high susceptibility genes to obesity has been also described in literature.

Two patients from western Spanish population, who were analyzed using array techniques in the context of their unaffiliated syndromic phenotype in newborns, resulted to be carriers of deletions including the *BDNF* and the *MC4R* gene, respectively. The exhibited phenotypes included severe features, whose severe clinical features compensated or at least delayed the overweight appearance. Weight and height were controlled, but no alterations were described at consultancy time. Each DNA sample was hybridized on the Affymetrix Cytogenetics Whole-Genome 2.7 Array. Data was collected using Gene Chip Scanner 3000 Dx and CEL files were analyzed using Chromosome Analysis Suite software (ChAS v1.1). The annotation file used was hg18 software was then used to reveal copy number at each locus quantifying the fluorescent signal intensity and comparing it with HapMap controls (International HapMap Project, 2005; http://www.hapmap.org).

Haploinsufficiency of the BDNF Gene

With respect to the example regarding to the haploinsufficiency of the *BDNF* gene, a girl who was affected with aniridia as the only clinical sign from birth, resulted carrier of the 40 Mb of 11p15.1p12 deletion [6]. High-density SNPs/CNVs microarray was used to delineate the deleted genes and to define how the area in haploinsufficiency could possibly contribute to the phenotype development. The proximal breakpoint was determined using the ChAS software at chromosome position 18,676,926 affecting the *TMEM86A* gene, and distal breakpoint at 36,576,388 within the *RAG2* gene (Fig. 3.1). The size deletion was 40 Mb encompassing a total of 124 genes and open reading frames, including the OMIM genes *CSRP3*, *SLC6A5*, *FANCF*, *BDNF*, *FSHB*, *PAX6*, *WT1*, *CD59*, *LMO2*, *CAT*, *CD44*, *RAG1* and *RAG2* (OMIM entrance ID: 601240). The predicted genotype-phenotype relationship was established from the published data of patients affected with WARG or WARGO syndrome, and those who were analyzed at older ages [7–9]. The patient carried a much broader than the described deletion which included all the major genes involved in the WARGO phenotype: *WT1*, *PAX6*, *BDNF* and *SLC1A2*. The sup-

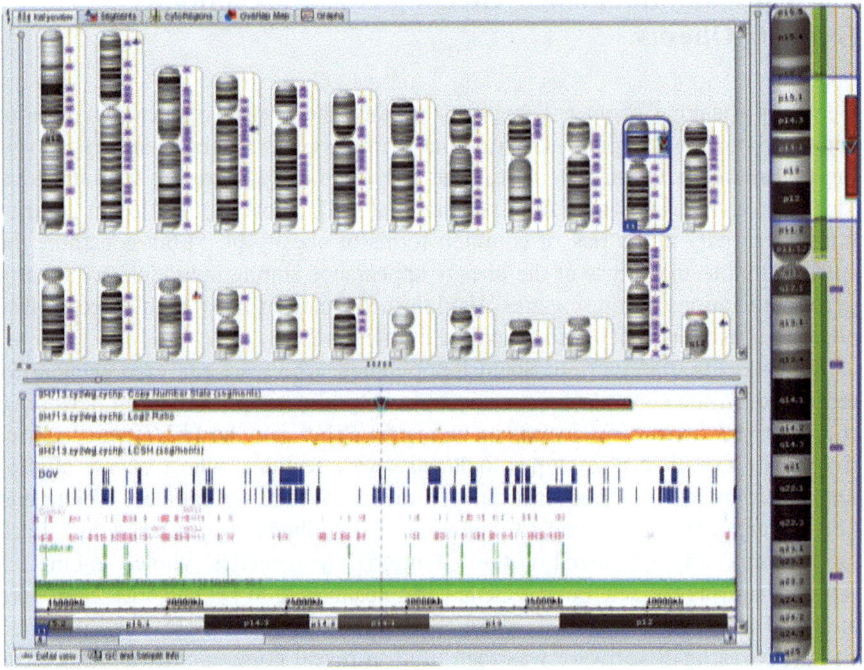

Fig. 3.1 Affymetrix Whole-Genome Human SNPs/CNVs 2.7 analysis. Deletion region 46,XXdel11p12-15.1 *dn* in patient (40 Mb). The distal breakpoint was determined using the ChAS software at chromosome position 18,676,926 affecting the *TMEM86A* gene and distal breakpoint at 36,576,388 within the *RAG2* gene. The *BDNF* gene is included in the deleted region

pression of *PAX6*, its regulatory elements located 5' of the gene, and other upstream elements explained the congenital aniridia and cataract in the patient. Sequencing of the remain allele of *BDNF* gene identified the Val66Met polymorphism, which has been associated with poorer episodic memory, abnormal activation of the hippocampus and eating disorders in distinct populations [10].

Rigorous monitoring of these patients would increase the clinical utility of the card for the syndrome genes WARG [11] and should define in detail the WARGO/WAGRO syndrome. In the Decipher database WARG 11p13 deletion syndrome is described to be associated to deletions variable in size (from 1 to 26.5 Mb) and always including the genes *WT1* and *PAX6*. Obesity is not included in the main features of the four described phenotypes, although is reflected that a substantial proportion of cases develop marked obesity, mediated by brain-derived neurotrophic factor (BDNF).

It has been described that the WARGO patients reach a BMI around 4.3 SD above the age-related population-based mean, at 8 years of age. By 10 years of age, 100 % of the patients with heterozygous *BDNF* deletions are obese (BMI > or = 95th centile for age and sex) as compared with 20 % of persons without *BDNF* deletions. The baby had not deviated from the predicted percentiles during the 1st year and a half of life, although the acceleration in weight gain has been described to start at the beginning of the second year of life [8].

Haploinsufficiency of the MC4R Gene

Other girl in our series (unpublished data), newborn, exhibited in physical examination a moderate axial hypotonia and dysmorphic features such as hypertelorism, flat nasal bridge, dysplastic ears, and redundant skin fold. The patient had an anorectal malformation consistent in a previous perineal fistula, with a previous ectopic anal opening in backplane fourchette [12]. At locomotor level, she had irreducible abducts and severe valgus feet. Cerebral ultrasound detected the cavity of septum pellucidum and vergae. The otoacoustic emissions and brainstem auditory evoked potentials were pathological, showing a conductive hearing loss. The temporary rock TAC showed stenosis of both ear canals. An echocardiogram was performed and an atrial septal defect (ASD), Ostium Secundum and patent ductus arteriosus (PDA) were diagnosed. Growth parameters were under the normal limits at testing time.

Data from Affymetrix Cytogenetics Whole-Genome 2.7 Array analysis evidenced two chromosomal misbalances: arr18p11.32p11.21 (136.226-12602631)×3, 18q21.32q23 (57691236-78014123)×1. 20.3 Mb subtelomeric deletion of 18q and 12.5 Mb duplication 18p subtelomeric (Fig. 3.2). Haploinsufficiency of the *MC4R* gene (Fig. 3.3) and the normal sequence at the remained allele by Sanger were corroborated. Clinical Follow up revealed that the girl maintained a ponder-estatural delay at 2 years of age, weight in percentile 1 (9 kg, corresponding to −2.52 SD) and the size also in P1 (80 cm, corresponding to −2.35 DE). The molecular bases of these signs could be related to the described growth hormone insufficiency in

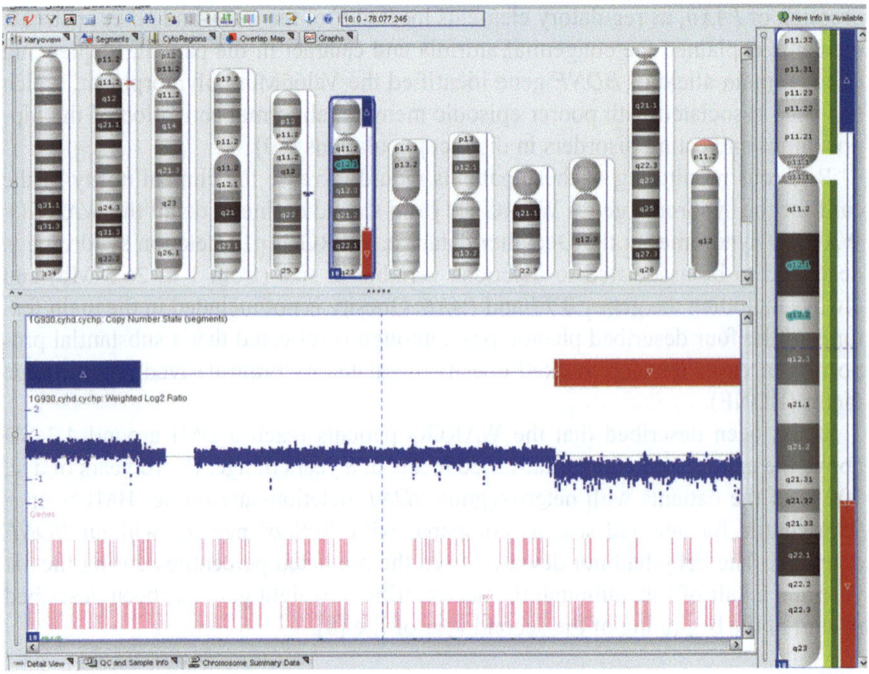

Fig. 3.2 Affymetrix Whole-Genome Human SNPs/CNVs CytoScan HD analysis. arr18p11.32p11.21 (136.226-12602631)x3, 18q21.32q23 (57691236-78014123)x1. 20.3 Mb subtelomeric deletion of 18q and duplication 18p subtelomeric 12.5 Mb. The *MC4R* gene is included in the deleted region

patients carrying deletions on 18q23 [13]. Further specifically designed studies of these patients will facilitate genotype-phenotype correlations for the haploinsufficiency of these specific genes responsible for the obesity. The comparison with published related phenotypes is essential [14]. It seems also essential maintaining a periodic clinical observance of these individuals to control the obesity indicators, and the most complete set of parameters and variables enable to influence them.

Phenotypic Correlation Between Loci, when Chromosomal Imbalances or Susceptibility Alleles to Obesity Exist

The genome-wide association studies (GWAS) demonstrated to be the most efficient study design in identifying common genetic variants associated with overweight. Using array techniques for case-control studies to the whole genome, the human obesity susceptibility was scanned with the maximum resolution levels. Around 250 obesity-related *loci* have been already identified [15–17] (Fig. 3.4). It

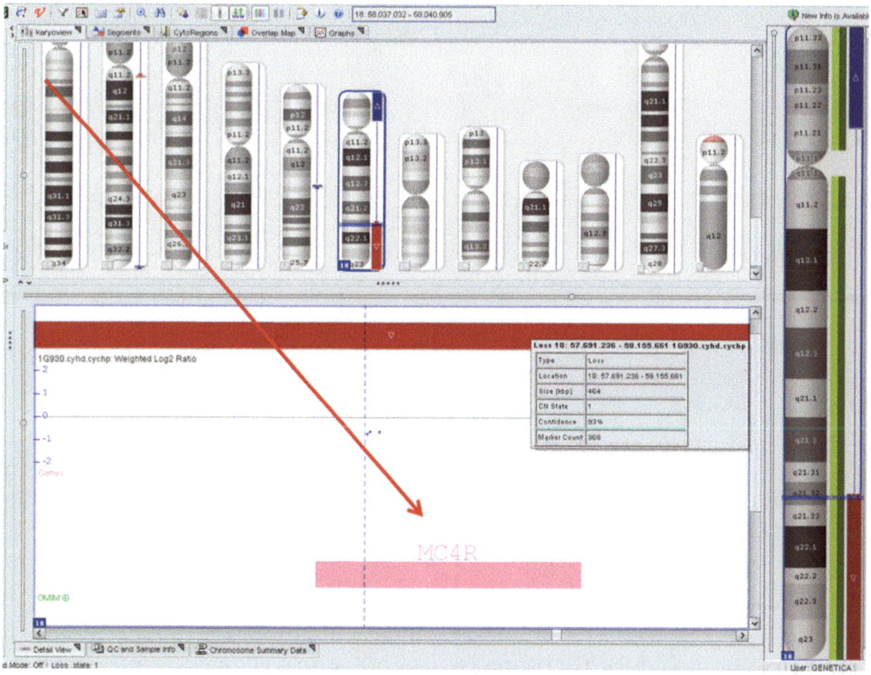

Fig. 3.3 Chromosome analysis suite software (Affymetrix) corroborated the existence of the happloinsufficiency of the MC4R gene, due to the 20.3 Mb subtelomeric deletion 18q21.32q23 (57691236-78014123)

has been clearly demonstrated the complexity of obesity, because many of these *loci* have been associated with BMI, other *loci* were associated with abdominal obesity, or the body fat percentage.

SNPs in *FTO* [18] and near-*MC4R loci* [19] were the locations firstly identified among Europeans as genetic location for obesity susceptibility. SNP-to-SNP comparisons suggested that more than half of the 36 body mass index-associated *loci* are shared across European and East Asian ancestry populations, whereas locus-wide analyses suggest that the transferability might be even more extensive [17].

The SNP rs9939609 in the *FTO* gene (Chr.16q12.2), has been one of the most extensively candidate variants studied, to which has been attributed to explain about 1 % of BMI heritability. Its common frequency allele and functional effect can be acting in an interesting percentage of obese individuals, whose characteristics classify their overweight as syndromic or not.

The information on Unique web page describes deletions which are vary in size with the proximal breakpoint at 16q12 to 16q13, including the *FTO* gene. Data is drawn from around 15 cases and the effects are extremely variable. Babies carrying deletions at 16q12 have been described to be born short and light, and continue to grow slowly, with their weight and height at or below the lowest curves on a growth chart. Obesity has not been referred in those patients. Descriptions of associated

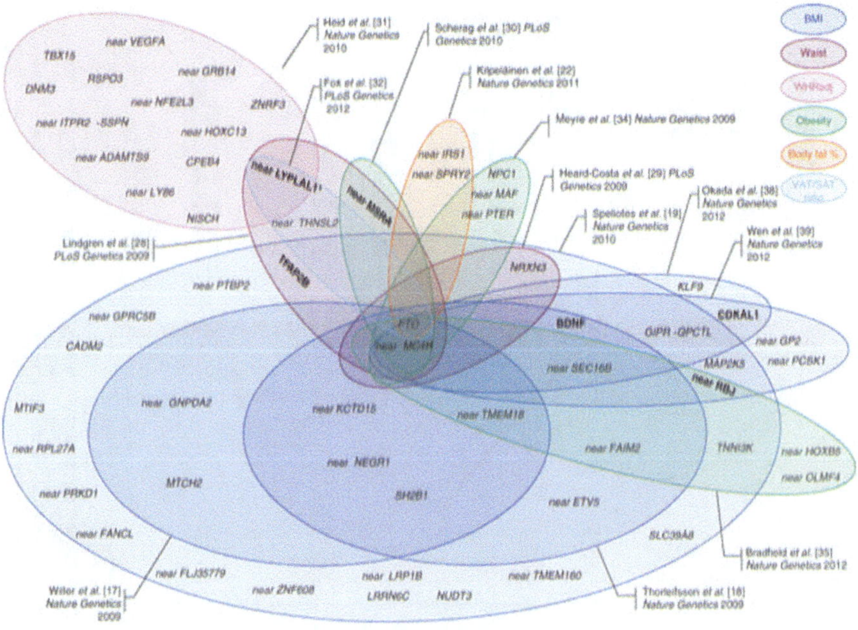

Fig. 3.4 Obesity-susceptibility loci identified through genome-wide association studies (GWAS) for body mass index (*blue*), waist circumference and waist-to-hip ratio (*pink*) and extreme and early onset of obesity (*green*). The fat mass and obesity associated gene FTO and the near-MC4R loci were corroborated as principal risk genomic regions. Figure modified and updated from Loos [17]

phenotypes in medical journals are usually episodic and frequently describe babies, with scant or no information on later health. 7 entries of 5 gains and 2 losses from 5.37 to 38.49 Mb have been detailed in Decipher v7.0 database, all of them included the *FTO* gene and none of their associated phenotype refers the existence of obesity.

Searching for individuals carrying deletions including the *MC4R* gene (Chromosome 18q21.32: 58,038,564−58,040,001) and flanking region, Unique web page describes cases whose height and build are above average. Children use to be on the 98th percentile for height. Obesity has been described among adults with deletions of 18q, who had grown up even plump but were frequently shorter than would have been expected. 24 entries of 11 gains and 13 losses from 2.17 to 28.15 Mb have been detailed in Decipher v7.0 database; all of them included the *MC4R* gene. The phenotype has been detailed in 12 cases from the 24 chromosomal imbalances described, and only the entry 218187 (deletion of 2.56 Mb) refers the existence of intellectual disability and truncal obesity (https://decipher.sanger.ac.uk/patient/248187).

The *NPC1* gene (Chromosome 18q11.2: 21,086,148−21,166,862) was identified as the gene that when mutated, results in Niemann-Pick disease type C (rare autosomal recessively inherited neurovisceral lipid storage disorder). Mutations in the *NPC1* gene have been strongly linked with obesity and genome-wide association studies, identified *NPC1* as a risk factor in childhood obesity (Fig. 3.4). Cases included in Unique database as carriers of deletions which include this genomic region, developed a severe overweight associated to behavioral problems and short stature. Decipher describes 7 entries of duplications which include the *NPC1* gene, the phenotypes have been detailed only in 3 cases. Obesity does not appear in the list of features.

One more example is the *SEC16B* gene (Chromosome 1q25.2: 177,893,091-177,953,438). Decipher web page details 14 entries, 11 losses and 3 gains, in which the *SEC16B* gene is implicated. 8 cases reported some data of their appearance; only two cases were detailed in deep. The entry 248552 (deletion of 20.95 Mb) described 36 distinct features of the exhibited phenotype, among them intellectual disability, truncal obesity and proportionate short stature (https://decipher.sanger.ac.uk/patient/248552). There is not information of cases carrying similar chromosomal misbalances in the Unique Rare Chromosome web page. It has been published the identification of this rare 1q25.2-31.3 deletion, using SNP-array analysis, spanning from 174,592,050 to 195,122,910 with a length of 20.5 Mb. The deletion was of paternal origin. Other genotypes and phenotypes of other 16 previously identified cases of 1q25-32 deletions including this genomic region, confirmed a similar phenotypes: obesity was not described as common features [20].

Copy Number Variants in Obesity-Related Syndromes

It has been corroborated than larger than 500 Kb and rare CNVs (< 1 %) contribute to the risk of common complex disorders, whose effects are less highly penetrant than Mendelian mutations [21]. Several rare, obesity risk CNVs have been detected by genome-wide studies of individuals with extreme phenotypes [22]. Among these, a 220 Kb deletion on chromosome 16p11.2 containing the *SH2B1* gene has been identified in three patients, who had inherited the deletion from obese parents. Also a proximal deletion of 600 Kb at chromosome 16p11.2 containing the *TBX6* gene was described in 2.9 % of series of obese children affected with cognitive deficits and/or congenital malformations [23]. Another study found eleven large and rare CNVs, which disrupt potential candidate genes as *UCP1* or *IL15* [24], contributing with higher risk to obesity than common variants.

The huge CNV dataset derived from the previously mentioned consortium of clinical laboratories (the International Standards for Cytogenomic Arrays consortium, ISCA; https://www.iscaconsortium.org), constitutes an efficient tool to research the impact of the existence of CNVs among thousands of patients with a variety of different clinical findings, compared with thousands of controls. These studies observe the frequency of obesity in individuals carrying specific deletions/duplications, highlighting the pathogenicity of the recurrent variants and its exact association to overweight.

These results served as a basis for a subsequent work in which 17 CNV *loci* were identified [25]. The deletion of the loci *EDIL3*, *S1PR5*, *FOXP2*, *TBCA*, *ABCB5* and *ZPLD1*, and the duplication of the *loci KIF2B* and *ARL*, replicated as risk factors to obesity susceptibility among distinct cohorts. A deletion at the *EPHA6-UNQ6114 locus* was also associated. Although these CNVs exhibited a low frequency among global population, contributes to the genetic susceptibility of common childhood obesity in subjects from different ethnicities. This work also detected that the *FTO* locus was very strongly associated with both BMI and obesity in their sample, however none association with any CNVs which was previously reported in subjects with severe pediatric obesity was observed.

Although rare CNVs with attributed risk to obesity, do not contribute greatly to the overall heritability for the disease, they constitute a potential insights into the underlying biology of childhood obesity; it is essential to mention that these variants constitute an important causal factor to their disorder.

Conclusion

This chapter determines the etiological screening of syndromic obesity, reflecting the place of array SNPs/CNVs, and detailing the recent advances in the field of genetics, mainly on the genes and genomic regions implicated in highly penetrant forms of obesity associated with developmental disorders. The efficiency of the array genomic techniques in this type of patients evidences a new field for investigating deleterious chromosomal imbalances and groups of CNVs that have not previously been detectable. Different syndromes with obesity have been associated with chromosomal imbalances found by arrays, but more studies in cohorts of patients selected for syndromic obesity are necessary. Their pathogenicity for obesity susceptibility has not been considered definitely proved in all patients. Designing individualized studies for these patients would facilitate to improve the knowledge about the genotype-phenotype correlations for specific genes that are implicated in the genetic variants, and that could be responsible for the obesity. The molecular characterization of cryptic chromosomal anomalies in these patients, could optimize and make much more rigorous their clinical management and follow-up, in addition to give the family the most adequate genetic counseling. The very low efficiency of the clinical management of extreme obesity when it is framed in a syndromic phenotype can only be improved if the patient care is related to the results of the clinical genetic testing.

Efforts of the ISCA model for incorporating phenotype data with genotype data, have generalized the use of the genome-wide analysis into clinical management on a broader scale. However, the clinical interpretation of these variants still remains difficult, particularly for those genes with recessive pattern, those, which have been partially disrupted, and those, which have not been well characterized. Improving the information related to the described phenotypes and the developed manifestations in the analyzed patients along the whole period life time is absolutely essential to define the deleterious capability of the novel variants and the genotype phenotype correlations.

Experts might design an efficient, precise and homogeneous way to aggregate the observed genotype and phenotype from the clinical testing and complete the data progressively, in order to improve the resources for clinicians and researches. Improving the valuable source of phenotypic data could redound within developing phenotypic profiles of emerging genomic disorders as obesity, the identification of candidate regions for that particular phenotype, and the creation of tools for use in clinical practice. The phenotypic description of new obesity-related syndromes will also point out a list of genomic regions that will help developing new-targeted diagnostic oligonucleotide arrays. A targeted clinical array to detect deletions or duplications occurring within genes in which the dosage effect is suspected to be cause the obese phenotype, could calculate a higher risk of inherited obesity.

In addition, the huge available data about human variation (SNPs and CNVs) have developed a new strategy in the search for genetic variants associated with complex phenotypes, using the Genome-wide linkage scans for identifying common loci associated with traits. However, the probe selection reflects the distribution of SNPs in the genome. The included SNPs were very sparse in genomic segments containing low-copy repeats and segmental duplications, and arrays frequently offer a better detection of CNVs in gene-poor regions. The following high-resolution SNP array platforms were generated by adding non-polymorphic sequences, covering the regions where SNP distribution is poorer.

It is essential to take into account that authors mentioned problems regarding GWAS based on SNPs/CNVs array analysis, specifically related to the inaccuracy of CNV calling algorithms in detecting particularly small CNVs, common CNVs, and duplications. Interesting articles insist on the CNV genotyping call errors on association analyses have not been extensively controlled and studied [26, 27]. The generated information will promote the combination of the new diagnostic arrays with massive sequencing platforms, which will enhance detection of known pathogenic genetic conditions. The interest to advance in common and expensive diseases for the health systems is developing a next-generation sequencing protocol relevant to clinical practice, also for obesity management.

References

1. Miller DT, Adam MP, Aradhya S, Biesecker LG, Brothman AR, Carter NP et al (2010) Consensus statement: chromosomal microarray is a first-tier clinical diagnostic test for individuals with developmental disabilities or congenital anomalies. Am J Hum Genet 86(5):749–764
2. Bell AC, Kremer PJ, Magarey AM, Swinburn BA (2005) Contribution of 'noncore' foods and beverages to the energy intake and weight status of Australian children. Eur J Clin Nutr 59:639–645
3. List JF, Habener JF (2003) Defective melanocortin 4 receptors in hyperphagia and morbid obesity. New Eng J Med 348:1160–1163
4. Rankinen T, Zuberi A, Chagnon YC et al (2006) The human obesity gene map: the 2005 update. Obesity (Silver Spring) 14:529–644

5. Saunders CL, Chiodini BD, Sham P et al (2007) Meta-analysis of genome-wide linkage studies in BMI and obesity. Obesity (Silver Spring) 15:2263–2275
6. Rodríguez-López R, Pérez JM, Balsera AM, Rodríguez GG, Moreno TH, García de Cáceres M et al (2013) The modifier effect of the BDNF gene in the phenotype of the WAGRO syndrome. Gene 516(2):285–290
7. Han JC, Liu QR, Jones M, Levinn RL, Menzie CM, Jefferson-George KS et al (2008) Brain-derived neurotrophic factor and obesity in the WAGR syndrome. N Engl J Med 359:918–927
8. Shinawi M, Sahoo T, Maranda B, Skinner SA, Skinner C, Chinault C et al (2011) 11p14.1 microdeletions associated with ADHD, autism, developmental delay, and obesity. Am J Med Genet A 155A(6):1272–1280
9. Xu S, Han JC, Morales A, Menzie CM, Williams K, Fan YS (2008) Characterization of 11p14-p12 deletion in WAGR syndrome by array CGH for identifying genes contributing to mental retardation and autism. Cytogenet Genome Res 122:181–187
10. Monteggia LM, Barrot M, Powell CM, Berton O, Galanis V, Gemelli T (2004) Essential role of brain-derived neurotrophic factor in adult hippocampal function. Proc Natl Acad Sci U S A 101:10827–10832
11. Clericuzio C, Hingorani M, Crolla JA, van Heyningen V, Verloes A (2011) Clinical utility gene card for: WAGR syndrome. Eur J Hum Genet 19(4), doi:10.1038/ejhg.2010.220
12. Schramm C, Draaken M, Bartels E, Boemers TM, Schmiedeke E, Grasshoff-Derr S et al (2011) De novo duplication of 18p11.21-18q12.1 in a female with anorectal malformation. Am J Med Genet Part A 155(2):445–449
13. Margarit E, Morales C, Rodríguez-Revenga L, Monne R, Badenas C, Soler A (2012) Familial 4.8 MB deletion on 18q23 associated with growth hormone insufficiency and phenotypic variability. Am J Med Genet Part 158A:611–616
14. Feenstra I, Vissers LE, Orsel M, van Kessel AG, Brunner HG, Veltman JA, van Ravenswaaij-Arts CMA (2007) Genotype-phenotype mapping of chromosome 18q deletions by high-resolution array CGH: an update of the phenotypic map. Am J Med Genet 143:1858–1867
15. Day FR, Loos RJ (2011) Developments in obesity genetics in the era of genome-wide association studies. J Nutrigenet Nutrigenomics 4:222–238
16. Loos RJ (2012) Genetic determinants of common obesity and their value in prediction. Best Pract Res Clin Endocrinol Metab 26:211–226
17. Lu Y, Loos RJ (2013) Obesity genomics: assessing the transferability of susceptibility *loci* across diverse populations. Genome Med 5(6):55
18. González JR, González-Carpio M, Hernández-Sáez R, Serrano V, Torres G, Rubio-Rodrigo M et al (2012) *FTO* risk haplotype among early onset and severe obesity cases in a population of Western Spain. Obesity (Silver Spring) 20(4):909–915
19. Willer CJ, Speliotes EK, Loos RJF, Li S, Lindgren CM, Heid IM et al (2009) Six new loci associated with body mass index highlight a neuronal influence on body weight regulation. Nature Genet 41:25–34
20. Hu P, Wang Y, Meng L, Qin L, Ma D, Yi L, Xu Z (2013) 1q25.2-q31.3 deletion in a female with mental retardation, clinodactyly, minor facial anomalies but no growth retardation. Mol Cytogenet 6(1):30
21. Gibson G (2011) Rare and common variants: twenty arguments. Nat Rev Genet 13:135–145
22. Bochukova EG, Huang N, Keogh J, Henning E, Purmann C, Blaszczyk K et al (2010) Large, rare chromosomal deletions associated with severe early-onset obesity. Nature 463(7281):666–670
23. Walters RG, Jacquemont S, Valsesia A, de Smith AJ, Martinet D, Andersson J et al (2010) A new highly penetrant form of obesity due to deletions on chromosome 16p11. 2. Nature 463:671–675
24. Wang K, Li WD, Glessner JT, Grant SFA, Hakonarson H, Price RA (2010) Large copy-number variations are enriched in cases with moderate to extreme obesity. Diabetes 59(10):2690–2694

25. Glessner JT, Bradfield JP, Wang K, Takahashi N, Zhang H, Sleiman PM et al (2010) A genome-wide study reveals copy number variants exclusive to childhood obesity cases. Am J Hum Genet 87(5):661–666
26. Zhang D, Qian Y, Akula N (2011) Accuracy of CNV detection from GWAS data. PLoS ONE 6(1):e14511
27. Wineinger NE, Tiwari HK (2012) The impact of errors in copy number variation detection algorithms on association results. PLoS ONE 7(4):e32396

Chapter 4
Genetic Contribution: Common Forms of Obesity

David Albuquerque and Clévio Nóbrega

What Is Obesity?

Human obesity is a global public health concern and results from an excessive accumulation of body fat that can adversely affect health [1]. The global rise of obesity has serious effects, and may contribute to a significant number of diseases including type two diabetes mellitus, cardiovascular diseases, metabolic syndrome, and some cancers [1, 2]. Excessive fat accumulation results from a persistent positive energy balance, where the amount of energy consumed exceeds the amount of energy expended [3]. So, a simple definition of obesity could be: a consequence of an imbalance between energy intake and energy expenditure [4]. The energy balance represents a conglomerate of traits, each one influenced by numerous variables such as behavior, diet, environment, social structures, metabolic factors and genetics [5]. The result of this complex interaction among all of these variables contributes to individual differences in the development of obesity.

Epidemiological studies indicate that adiposity, as reflected by BMI, has increased worldwide over the past decades [6]. Moreover, obesity is more common in some countries than in others, though precise cross-country comparisons can be difficult because not all samples are representative of the relevant populations. Nonetheless, available data suggest that the increase in the prevalence of obesity began to emerge during the 1980s and ever since more countries have joined the global obesity pandemic [6]. In modern societies, despite obesity awareness campaigns

D. Albuquerque (✉)
Faculdade de Ciências e Tecnologia, University of Coimbra, Coimbra, Portugal
e-mail: dav.albuquerque@gmail.com

C. Nóbrega
Center for Neurosciences and Cell Biology, University of Coimbra,
3004–517 Coimbra, Portugal
e-mail: clevio@cnc.uc.pt

C. Nóbrega, R. Rodriguez-López (eds.), *Molecular Mechanisms Underpinning the Development of Obesity*, DOI 10.1007/978-3-319-12766-8_4

and efforts to decrease the energy intake and increase the energy expenditure, obesity prevalence is still increasing. Obesity has a multi-factorial etiology, involving various non-genetic and genetic factors [1, 7]. Probably, most cases of obesity results of a cluster towards the middle of this spectrum, which can be best described as the outcome of an adverse obesogenic environment, working on a susceptibility genotype. Effectively, the genetic susceptibility can potentially be mediated through defects in several different homeostatic mechanisms. Certainly, the exposure to an obesogenic environment could be the cause of the increase in the prevalence of high BMI for the last 30 years [1, 2, 6].

The Genetics of Human Obesity

The increase in the obesity prevalence around the world has been broadly attributed to the change in environment, which is more obesogenic. On the other hand, specific features of the energy balance mechanisms can effectively protect against obesity, possibly explaining why one third or more of the population remains lean [8]. The obesity phenotype only emerges if food consumption exceeds the energy expenditure on a lasting basis, resulting in a prolonged positive energy balance. However, there are many risk factors that predict the development of obesity and generally all involve the interaction of biological and social factors. Numerous studies are consistent with the thesis that the personal genetic profile could be a cause for individual differences in the predisposition to weight gain. It is, therefore, interesting that most of the genes involved in the susceptibility of obesity are also related to food intake and regulation of energy balance [8].

The Evidence for a Genetic Component in Obesity

Over the last 30 years, the increase in the prevalence of obesity could be attributed to environmental changes, or to high-calorie food intake together with the sedentary lifestyle of modern societies [7]. However, these data appear to be inconsistent with twin and adoption studies that have concluded that approximately 70 % of the variance in obesity is due to genetics [9–11]. The fact that the prevalence of obesity in many countries has increased three-fold over the last three decades seems incompatible with the notion that genetics are the primary cause of obesity. Thus, other factors, like methylation (or other epigenetics mechanisms), in which environmental factors cause changes in the expression of genes, could explain the increase in obesity prevalence.

Heritability represents the proportion of phenotypic variation among individuals due to genetic contribution. Hence, it is not surprising that one important risk factor for childhood and adolescent obesity is parental obesity. Whitaker et al. [12] found that when both parents are obese there is an increase of more than double of the risk for childhood obesity. However, most of the studies found a small to medium effect of parental obesity as risk factor for childhood obesity [13]. Other studies have found a stronger effect for maternal obesity compared to paternal obesity, which may reflect pre- and postnatal environmental factors [14]. Moreover, maternal weight gain in pregnancy has been positively associated with BMI of the children into adulthood [15].

Twin studies have been used to model the genetic component of a given trait, due to the fact that monozygotic (MZ) twins are genetically identical, while non-identical dizygotic (DZ) twins share only 50 % of their genetic material [7]. In 1977, Feinleib et al. [11] studied the correlations for weight in 250 MZ and 264 DZ male veteran twin pairs, and established for the first time that familial aggregation for obesity results mainly from genetic influence. In 1986, Stunkard et al. [10] confirmed these results in a 25-year follow-up study using more than 4000 MZ and DZ twin pairs. High heritability values for BMI were observed for the same subjects at 20 years ($h^2=0.77$) and at 45 years ($h^2=0.84$). The heritability of fat mass among MZ twins has been reported to range from 70–90 %, while in DZ twins it is 35–45 %. Adoption studies have strengthened the evidence of a strong genetic influence on human body weight. Body corpulence of adopted children correlates more strongly with BMI of their biologic parents *versus* the BMI of their adoptive parents [9]. Recently, Silventoinen et al. [3], conduct a review of studies in twins and adopted children, suggesting that genetic factors could have a much stronger effect than environmental factors on the BMI trends in children up to the age of 18 years.

Another evidence for a genetic component in obesity is highlighted through the different prevalence between racial groups. For example, it was found that obesity prevalence in Caucasian and Asian populations is about 35 % or less compared to 50 % or more found among Pima Indians living in New Mexico [16]. Several studies support the concept that genes play a key role in the obesity etiology. However, the search for underlying genotypes that cause of obesity has been challenging due to the complex interactions involved in the regulation of adiposity. Indeed, the vast majority of individual genotypes that have been associated with elevated body mass have not been replicated in a reliable fashion. Moreover, environmental factors and cultural diversity also account for the different obesity prevalence found across ethnicities.

Polygenic or Common Forms of Obesity

In most modern societies, the environment favors weight gain due to food abundance and lack of physical activity, favoring an increase of common obesity in both adults and children worldwide. The genetic profile of common obesity results from the effects of several altered genes [17]. However, the genetic and molecular mechanisms involved in body weight regulation are complex and not yet completely understood. In theory, the genetic basis of common obesity implies that the specific set of variants relevant for obesity vary considerably from one obese person to the next [18].

The Genetic Approach for Common Obesity

The study of common obesity is based in the analysis of gene variation in genomic DNA (single nucleotide polymorphism, or repetition of bases of polyCAs or microsatellites) situated within or near candidate genes. In contrast with monogenic obesity, in polygenic obesity each mutation leads to a variant that confers susceptibility, requiring additionally the presence of other variants and an obesogenic environment to determine the obese phenotype [19]. There are three main approaches for the detection and analysis of a candidate gene in body weight regulation: linkage studies, candidate gene association studies and GWA studies. Their objective is to determine whether an association between a genetic variation and obesity-related trait exist.

Linkage Studies for Common Obesity

Family-based genome-wide linkage scans attempts to identify chromosomal regions co-segregating with disease-related phenotypes of interest in related individuals. This design was used in many studies that successfully localized the causes of rare monogenic disease, being frequent for the study of complex diseases. Until 2006, over 250 quantitative trait *locus* (QTLs) in more than 60 genome scans have been reported to be associated with adiposity in every chromosome except the Y chromosome [17]. Using the positional cloning strategy based on linkage, results have identified candidate genes, including glutamic acid decarboxylase enzyme (*GAD2*) [20], ectonucleotide pyrophosphatase/phosphodiesterase 1 (*ENPP1*) [21] and solute carrier family 6 (amino acid transporter), member 14 (*SLC6A14*) [22] directly related to obesity.

Candidate Gene Study

The candidate-genes association study approach detects association between genetic markers in pre-selection genes considered to be candidate for the studied phenotype. In obesity studies the focus has been on polymorphisms in a gene within a linkage region [23] or genes known for coding proteins involved in the regulation of lipid and glucose metabolism, food intake or energy expenditure [24, 25]. More than 300 chromosomal *loci* show some evidence of linkage with obesity [26, 27]. Nevertheless, few genes (~20) supported by at least five positive studies have been associated with common obesity using the candidate gene approach [28]. One of the main limitations of this approach is the need to understand the biological pathways of the disease. Generally, most of the complex diseases like obesity are not yet understood, nor the molecular mechanisms underlying its pathogenesis. Interestingly, genetic mutations that produce monogenic forms of obesity do not appear to be involved in the development of common forms of obesity [28].

Genome-Wide Approach

The first results about human variation provided by the Human Genome Project (initiated in 1990) led to the development of a new strategy in the search for genes associated with complex phenotypes [4]. The Genome-wide linkage scans identify *loci* associated with complex traits. This strategy involves the typing of families with multiple cases and tests some chromosomal regions capable of harboring one or more genes co-segregating with the trait. However, in the case of obesity, these studies did not identify multiple variants, mostly because common variants with high penetrance do not contribute substantially to the risk of common forms of obesity [25].

By 2006, the advance in chip genotyping technology conducted by the International HapMap project enabled the development of a new approach to investigate the genetic basis of complex diseases [23, 24]. The GWAS proved to be powerful and efficient in the identification of genetic variants associated with complex diseases [29]. Using many common variants, this case-control study approach screened the whole genome at much higher resolution than the approach previously described. This method operates in two stages: the first is the discovery stage in which millions of polymorphisms across the genome are tested for association with a particular trait, and then, the second stage consisting in the identification of polymorphism, which are associated with the trait, and subsequent tests of the presence of these polymorphisms in a new population [30]. It is now possible to identify several common polymorphisms associated with a particular trait, and like for other complex diseases, the advent of GWAS permitted to identify several *loci* associated with obesity [24, 27].

New Approaches for Common Obesity

The advent of automated DNA sequencing instruments, involving advances in engineering, chemistry, molecular biology, and software, opened a huge number of new opportunities [31]. Currently, molecular diagnosis based on Sanger's sequencing is restricted to only a few genes as this technology is expensive, time consuming, and labor intensive. The advent of next-generation sequencing (NGS) technology provides a new method for molecular diagnosis, which consists in the identification of genetic variations within several genes at the same time [30], promising to change the landscape of genetic testing with innovative cost-efficient methods for sensitive obesity multi-gene screening.

Some studies can be found using the NGS technology to study obesity. Saeed et al. [32] analyzed 26 susceptible genes for obesity in a sample of 39 Pakistani children with early-onset obesity. They found two new *LEPR* mutations at the homozygous state: a splice site mutation in exon 15 (c.2396−1 G>T), and a nonsense mutation in exon 10 (c.1675 G>A). Sällman et al. [33] amplified the entire region of *FTO* gene (412 kilo base pairs), from 524 severely obese and 527 lean Swedish children. They detected 705 single nucleotide polymorphisms, from which 19 were novel obesity-associated polymorphisms within the first intron of the *FTO* gene. An interesting finding was the fact that ten of them have a stronger association with obesity ($p<0.007$) when comparing with the commonly studied rs9399609 polymorphism ($p<0.012$). This study concluded that within the entire region of the *FTO* gene the first intron was the only one associated with obesity. Saeed et al. [32] searched for mutations with NGS in 40 patients with a monogenic form of diabetes ($n=19$) or obesity ($n=21$), in which the causing mutation was already known. The study found the same mutations described as the phenotype cause, except for one variant (mean of 98.6 %). On the other hand novel mutations were found in three patients with a putative deleterious effect.

The NGS approach could be used as an efficient tool, highly sensitive for the screening of mutations in genes associated with obesity or other diseases. Further, sequencing the human genome can now be accomplished in the data-generation phase within two weeks at a cost of approximately US$ 5000 [31]. However, the price for genome sequencing continues to decrease; in 2014, Illumina announced that will produce a new system called HiSeq X Ten that can deliver full coverage of human genomes for less than US$ 1000.

Common *Loci* Associated With Obesity-Susceptibility Discovered Through GWAS

The GWAS approach has been the most common methodology used, allowing geneticists to scan numerous polymorphisms (~0.1–5 million of polymorphisms) across the entire genome using powerful statistical methods to identify *loci* associated with a particular phenotype. Since the start of the GWAS era in 2005, there have been five waves of GWAS' discoveries for BMI (Table 4.1). The first *loci* identified through GWAS was the fat mass and obesity-associated (*FTO*) gene, and currently more than 50 genetic *loci* have been identified as being associated with at least one obesity-related traits [4, 7, 27].

Table 4.1 Currently established *loci* associated with BMI in GWAS

Wave	Gene symbol	Gene Name	SNP ID	Effect size BMI (OR 95%CI)[a]	Discovery study
First	*FTO*	Fat mass and obesity associated	rs9939609	1.31 (1.23–1.39)	[35, 46]
Second	*Near MC4R*	Melanocortin-4 receptor	rs17782313	1.12 (1.08–1.16)	[61]
Third	*Near TMEM18*	Transmembrane protein 18	rs7561317	1.20 (1.13–1.27)	[66]
			rs6548238	1.19 (1.10–1.26)	[64]
	FAIM2	Fas apoptotic inhibitory molecule 2	rs7138803	1.14 (1.09–1.19)	
	Near GNPDA2	Glucosamine-6-phosphate deaminase 2	rs10938397	1.12 (1.07–1.17)	
	SEC16B	S. cerevisiae Sec16	rs10913469	1.11 (1.05–1.18)	
	BDNF	Homolog of brain-derived neurotrophic factor	rs925946	1.11 (1.05–1.16)	
	Near ETV5	Ets variant 5	rs7647305	1.11 (1.05–1.17)	
	SH2B1	SH2B adaptor protein 1	rs7498665	1.11 (1.06–1.17)	
	Near NEGR1	Neuronal growth regulator 1	rs2568958	1.07 (1.02–1.12)	
	Near KCTD15	Potassium channel tetramerization domain containing 15	rs29941	1.10 (1.04–1.15)	
			rs11084753	1.04 (0.98–1.10)	
	MTCH2	Mitochondrial carrier 2	rs10838738	1.03 (0.98–1.08)	

Table 4.1 (continued)

Wave	Gene symbol	Gene Name	SNP ID	Effect size BMI (OR 95%CI)[a]	Discovery study
Fourth	*Near PRKD1*	Protein kinase D1	rs11847697	1.10 (1.03–1.17)	[67]
	SLC39A8	Solute carrier family 39, member 8	rs13107325	1.10 (1.05–1.15)	
	TFAP2B	Transcription factor AP-2 beta	rs987237	1.09 (1.05–1.12)	
	QPCTL	Glutaminyl-peptide cyclotransferase-like	rs2287019	1.09 (1.05–1.12)	
	NRXN3	neurexin 3	rs10150332	1.09 (1.05–1.12)	
	Near GPRC5B	G protein-coupled receptor, family C, group 5, member B	rs12444979	1.08 (1.04–1.11)	
	Near RBJ-DNAJC27	DnaJ (Hsp40) homolog, subfamily C, member 27	rs713586	1.07 (1.05–1.09)	
	MAP2K5	Mitogen-activated protein kinase 5	rs2241423	1.07 (1.04–1.10)	
	Near TMEM160	Transmembrane protein 160	rs3810291	1.06 (1.03–1.08)	
	Near FANCL	Fanconi anemia, complementation group l	rs887912	1.06 (1.03–1.08)	
	Near FLJ35779-POC5	Centriolar protein	rs2112347	1.05 (1.03–1.08)	
	NearLRP1B	Low density lipoprotein receptor-related protein 1b	rs2890652	1.05 (1.02–1.08)	
	MTIF3	Mitochondrial translational initiation factor 3	rs4771122	1.05 (1.01–1.08)	
	LRRN6C	Leucine rich repeat neuronal 6c	rs10968576	1.04 (1.02–1.06)	
	TNNI3K	Interacting kinase	rs1514175	1.04 (1.02–1.07)	
	CADM2	Cell adhesion molecule 2	rs13078807	1.03 (1.00–1.06)	
	NUDT3	Nucleoside diphosphate linked moiety x type motif 3	rs206936	1.03 (1.01–1.06)	
	Near RPL27A	Ribosomal protein l27a	rs4929949	1.03 (1.01–1.05)	
	Near ZNF608	Izinc finger protein 608	rs4836133	1.03 (1.01–1.05)	
	Near PTBP2	Polypyrimidine tract binding protein 2	rs1555543	1.02 (0.99–1.04)	

Table 4.1 (continued)

Wave	Gene symbol	Gene Name	SNP ID	Effect size BMI (OR 95%CI)[a]	Discovery study
Fifth	*GNAT2*	Guanine nucleotide binding protein (g protein) alpha transducing activity	rs17024258	1.27 ($p = 0.02$)	[68][b]
	HS6ST3	Heparin sulphate 6-o-sulfotransferase 3	rs7989336	1.09 ($p=0.0001$)	
	HNF4G	Hepatocyte nuclear factor 4, gamma	rs4735692	1.09 ($p=1.97\times10\text{-}5$)	
	RPTOR	Regulatory associated protein of mtor, complex 1	rs7503807	1.08 ($p = 7.07\times10\text{-}5$)	
	MRPS33P4	Mitochondrial ribosomal protein s33 pseudogene 4	rs13041126	1.08 ($p = 0.001$)	
	ZZZ3	Zinc finger, zz-type containing 3	rs17381664	1.08 ($p=0.001$)	
	ADCY9	Adenylate cyclise 9	rs2531995	1.06 ($p=0.01$)	

BMI body mass index, *OR* odd ratio; 95 % *CI* confidence interval, *SNP ID* polymorphism identification

[a] Effect size from first discovery study

[b] This study this not reported confidence intervals, but rather *p*-values

First Discoveries by GWAS: *FTO* Gene

The first *locus* to be associated with obesity was the insulin-induced gene 2 (*INSIG2*) [34]. However, replication studies demonstrated very inconsistent results. So, the first *locus* unequivocally associated with obesity by a GWA study was the *FTO* gene [35]. Initially, Frayling et al. [35] conducted a GWA study to test the correlation between polymorphisms across the entire human genome and type II diabetes (T2D). They found that the rs9939609 polymorphism, located in the first intron of the *FTO* gene was strongly associated with T2D and increased BMI. However, after adjustment for BMI, the apparent association of the polymorphism with T2D was not maintained. The effect size of *FTO* polymorphism on BMI is modest, with homozygous individuals for the risk allele (in this case a "A") weighing on average 3 kg more than those homozygous for the protective allele (in this case a "T"), with the difference representing approximately 0.36 kg/m^2 [7]. These findings have been independently replicated and have consistently confirmed the association of rs9939609 polymorphism with the etiology of common obesity in several populations: European [36, 37], Asian [38–41] and African [42–44], both in children and adults. Two following studies reporting other polymorphisms in the intronic *FTO* region were also consistently associated with severe early-onset childhood and adult obesity (rs1421085 and rs17817449) [45], and have extended the association to other obesity-related traits including body weight and waist-to-hip circumference ratio (WHR) (rs9930506) [46]. The *FTO* polymorphisms were also associated with abdominal obesity, waist circumference and waist-to-hip ratio (WHR) [47, 48], and also with body fat percentage [49]. Although the findings replicate well, the *FTO* polymorphisms explain only 1–3 % of the variance in BMI [35, 46].

The functional mechanism underlying *FTO* role in obesity remains unknown, as well as the pathway underlying that role. The *FTO* is a very large gene with 9 exons spanning more than 400 kb in the chromosome 16q12.2 [50]. It was originally identified in 1999 in the mouse Fused toes (*Ft*) homologue mutant, resulting in a deletion of 1.6 Mb on chromosome 8 [51]. Homozygosity of *Ft* mutants is embryonically lethal. To investigate the biological function of *FTO* gene, two mouse models were used. Homozygous *FTO−/−* mice introduced by Fischer et al. [52] show postnatal growth retardation, significant reduction in fat and lean body mass compared to the wild-type animals [53]. In other mice model it was observed a lean phenotype in mice carrying a missense mutation in exon 6 of *FTO* (*FTO* I367F mice) [54]. These results seem to indicate that *FTO* could play a role in food intake control, energy expenditure and homeostasis.

The predicted human protein consists of 505 amino acids, characterized as a 2-oxoglutarate-dependent enzyme that is localized in the cell nucleus, belonging to the (2OG) oxygenases AlkB family of proteins [55]. The AlkB is a DNA repair enzyme, which catalyzes Fe(II)- and 2OG-dependent demethylation of damaged DNA substrates [56]. Recently, a study indicated that *FTO* also demethylates N6-methyladenosine (m6 A) residue in nuclear RNA [57]. *FTO* variation appears to lead to an increase in energy intake [58] by modifying hypothalamic control of appetite [59]. The crystal structure of *FTO* has recently been published and reveals the basis for its substrate specificity [60].

To date, over 500 studies have been performed concerning the association of *FTO* polymorphisms with obesity in several populations worldwide, and more than 60 polymorphisms in this gene were significantly associated with obesity [59]. All these polymorphisms were found within a 47 kb linkage disequilibrium (LD) block encompassing parts of the first two introns as well as exon two of *FTO* gene [61]. This is a region where the sequence is strongly conserved across species, and were polymorphisms are highly correlated (LD $r^2 > 0.80$ in CEU of the HapMap) in Caucasian populations [59].

Five Waves of GWAS

Following the discovery of the *FTO locus*, investigators enhanced GWA studies by increasing the sample size improving the statistical power to uncover additional obesity-susceptibility *loci*. Subsequently, a large-scale international consortium, called the Genetic Investigation of Anthropometric Traits (GIANT) emerged. The association data of 16.876 Caucasians from seven GWAS for BMI were combined in a meta-analysis [62]. This study confirmed the strong association of obesity with polymorphisms in the *FTO* gene, and identified one new *locus* near the *MC4R* gene which mutations are known to be the common cause of extreme childhood obesity [62]. The *MC4R* was the second gene significantly associated with common obesity [62, 62]. The rs17782313 polymorphism near the *MC4R* gene was associated with obesity among both adults and children [63]. Another polymorphism (rs12970134) near the *MC4R* gene also appears to increase the risk of obesity among Europeans [65]. Several polymorphisms near the *MC4R* gene have subsequently been found and replicated in various populations of European descents, as well as in Asians [66], African American [66], and in children and adolescents [42, 44].

In the third wave of discoveries, a meta-analysis was performed using 15 GWAS for BMI in Caucasians ($n > 32.000$) and replicated in another 14 studies for a second-stage sample of 59.082 individuals [67]. They confirmed the association of the *FTO* and *MC4R* genes, and found six new genes positively associated with obesity: *MTCH2*, *GNPDA2*, *KCTD15*, *SH2B1*, *NEGR1* and *TMEM18*. At the same time, a GWAS of 31.392 individuals, predominantly from Iceland population, found seven new genetic *Loci* near or in: *BDNF*, *SEC16B*, *ETV5* and *FAIM2*, as well as *FTO* and *MC4R* genes associated with BMI [65]. Four of the seven newly identified *loci* were common with the results from Willer et al. [67].

In 2010, the fourth wave, the GIANT consortium expanded its GWAS stage to comprise 249.796 individuals of European origin, and reveal 18 new *loci* associated with BMI near or in: *PRKD1*, *SLC39A8*, *GPRC5B*, *MAP2K5*, *QPCTL*, *RBJ*, *LRRN6 C*, FLJ35779, *CADM2*, *TMEM160*, *FANCL*, *LRP1B*, *TNNI3K*, *MTIF3*, TFAP2B, *ZNF608*, NRXN3, *RPL27A*, *PTBP2* and *NUDT3* [68]. By 2011, GWAS had identified 32 genetic *loci* unequivocally associated with BMI.

The most recent and fifth wave expanded the GIANT meta-analysis, to comprise 263.407 individuals of European ancestry [69]. Besides confirming all 32 BMI-associated *loci* previously identified by the fourth wave, they found seven new *loci*: *ZZZ3*, *RPTOR*, *ADCY9*, *GNAT2*, *MRPS33P4*, *HS6ST3* and *HNF4G*, explaining an additional 0.09 % of the variability in BMI [69].

To date, more than 35 *loci* have been found associated with the increase of BMI (explaining ~ 1–2 % of the variance in BMI), while other *loci* correlate with abdominal obesity, establishing 13 *loci* associated with it, assessed by the WHR [70]. Other *loci*, such as the Lactase gene (*LCT*) have been associated with BMI and abdominal obesity, but more studies are required to confirm associations [71–74]. A study identified two new *loci* with body fat percentage: *IRS1* and the other near *SPRY2* [75]. There is a gap between explained variance due to known common polymorphisms that explain 1–2 % and the estimated heritability of BMI (40–70 %). One of the main problems pointed out in GWAS is the failure to detect *loci* that are associated with traits whose effect sizes are too small to reach genome-wide statistical significance (false negative rate). To circumvent this "missing heritability" the genome-wide complex trait analysis (GCTA) method appears to show a multitude of low penetrance common polymorphisms, each with causal effects but too small to allow detection by GWA studies. Using this approach, Yang et al. [76] estimate the genetic variation for BMI to 17 % and in a recent analysis of twin studies revealed that 37 % of BMI could be explained by additive effects of multiple common polymorphisms [77]. Finally, a recent study found that BMI-associated *FTO* variants interact with the promoter region of iroquois homeobox 3 (*IRX3*) gene in the human, mouse and zebrafish genomes [78]. They also found that in Irx3-deficient mice, there is a reduction in body weight of 25–30 %. However, the *IRX3* gene had not been previously identified as associated with BMI in a GWA study. All these data confirmed the complexity of the genetics underlying obesity.

Testing Adult-Discovered *Loci* in Children

Childhood obesity is a major health problem in developing countries throughout the world. Most of obesity susceptible genes were found in studies with adults, which prompted an effort to replicate findings in studies with children [36, 44, 79, 80]. Knowledge of the genetic risk factors of obesity in children could be used as a first step to develop possible prevention measures. The *FTO locus* remains the most replicated gene and the strongest gene associated with obesity susceptibility, both in adults and children [36, 44, 81]. Genes *TMEM18* and *GNPDA2* were also associated with obesity susceptibility, with a similar effect of the *FTO* gene [82]. The remaining *loci* with evidence for association were *INSIG2*, *MC4R*, *NEGR1*, *BDNF* and *KCTD15* [82, 83].

In the GIANT meta-analysis of adult BMI in a pediatric European American sample, Zhao et al. [80] examined 32 genetic *loci* in 1.097 obese cases and 2.760 lean controls, aged between 2 and 18 years old. They found evidence of associations with nine of these *loci*, namely at *FTO*, *TMEM18*, *NRXN3*, *MC4R*, *SEC16B*, *GNPDA2*, *TNNI3K*, *QPCTL*, and *BDNF*. Overall, 28 of the 32 *loci* showed directionally consistent effects to that of the adult BMI meta-analysis.

Another similar report by the Early Growth Genetics (EGG) consortium investigated the effect of established adult BMI with two recently associated *loci* with childhood obesity (*HOXB5* and *OLFM4* genes) [84] in a Greek adolescents cohort [85]. The genetic risk score of the 34 (GRS-34) variants was calculated and found that variants at the *FTO*, *TMEM18*, *FAIM2*, *RBJ*, *ZNF608* and *QPCTL loci* produced nominal evidence for association with BMI and/or obesity risk. Overall, 27 out 34 variants showed consistent effects with those reported by large-scale meta-analyses adult BMI.

These results showed clearly that these obesity-conferring variants operate early in life, suggesting that individual preventative lifestyle intervention in childhood could be important to obesity development.

GWAS-Related Investigations in Other Ethnicities

There are remarkable disparities in the prevalence of obesity between ethnic groups. To date most of GWAS reports published have been performed in populations of European origin. Only one study identified, a *locus* near *MC4R* gene associated with waist circumference and insulin resistance in a cohort of South Asian population [64]. This could be partly due to the fact that some susceptible *loci* only affect a specific ethnic group, while others might affect other ethnic group. Indeed, the human genetic architecture differs across ethnicities, which is well illustrated by differences in linkage disequilibrium (LD), whereas haplotype blocks vary only somewhat among human populations [86].

As a case in point, *FTO locus* have consistently correlated with BMI and risk of obesity in populations of African [42–44, 87], Asian [38–41] and Pacific-Islander [88] ancestry. Despite the fact that effect sizes were similar to those observed in white European populations, the risk of allele frequency varies substantially: ~25 % in Asian, ~45 % in white Europeans and range of~ 7–18 % in African origin [89].

Two recent independent meta-analysis were performed in both East Asians and African populations [90, 91]: Wen et al. [90] performed a meta-analysis using 27.715 individuals, followed by *in silico* and *de novo* replication studies in a further 37.691 and 17.642 individuals of East Asians, respectively. Seven previously identified *loci* were detected (*FTO*, *SEC16B*, *MC4R*, *GIPR-QPCTL*, *ADCY3-RBJ*, *BDNF* and *MAP2K5*) and three new *loci* were uncovered, near or in *CDKAL1*, *PCSK1* and *GP2* genes. Data also implicated three *loci*, *GNPDA2*, *TFAP2B* (previously identified) and *PAX6*, which all reached the genome-wide significance threshold. A recent meta-analysis was conducted to examine the association of >3.2 million polymorphisms with BMI in 39.144 adults of African ancestry [91]. It identified one new *locus* at 5q33 (*GALNT10*, rs7708584 polymorphism) and another at 7p15, when data from the GIANT consortium was included (*MIR148A-NFE2L3*, rs10261878 polymorphism). They also found evidence of an association at 6q16 (*KLHL32*, rs974417 polymorphism) in African-ancestry sample. Overall, 32 of the 36 previously established BMI variants showed consistent effect in this GWAS. The 36 known BMI *loci* explain in average 1.30 % of the variance in BMI of African ancestry compared with 1.67 % and 1.25 % in European and Asian ancestry populations, respectively [91]. More recently, Tan et al. [91] replicated 6 confirmed obesity genes (*FTO, CTNNBL1, ADRB2, LEPR, PPARG* and *UCP2* genes) in 8 different samples from different ancestries (five Caucasian, one Chinese, one African-American and one Hispanic population). Regarding only the *FTO* gene they found 35 polymorphisms significantly associated with obesity in Caucasian population. However, all of them showed limited or no evidence of associations with obesity in the other ethnic groups.

Association studies across different populations can help us to define more precisely which *loci* or variants could play a role in the obesity etiology, and help to understand the genetic and environmental factors that could contribute to obesity. The discovery of new *loci* in replication studies at established *loci* found in other populations reflects differences in allele frequency and effect size. Further studies will be needed to test the biological function at the associated *loci*.

The Obesity Risk-Allele Scores

As noted, several GWAS have identified a large number of obesity susceptibility *loci*. Nevertheless, the major part of these studies only identified single genetic *loci* associated with obesity. It has been demonstrated that combining information from all these obesity *loci* into a genetic risk-allele scores (GRS) could be a convenient way to summarize a risk-associated variation across the genome [93] and better when individual genetic effects are moderate [94]. The simplest way to calculate a GRS is by summing the number of accumulated risk alleles associated with the disease.

The use of a combined genetic score is considered as a better tool to determine the susceptibility of a common trait, than using each genetic *locus* alone. This is particularly more evident when the allele score consists either of many common polymorphisms with small effects, or of rare polymorphisms [94]. Generally, when several polymorphisms are combined into the same allele score, the score may explain a considerable proportion of variation in the risk factor, even if none of the polymorphisms individually does. In complex diseases it is likely that the effects of different genetic *loci* related to obesity operate in an interactive fashion. Future research should investigate this possibility using classification or regression tree analyses, which are well suited to detecting complex non-linear interactions. The identification of the complex interplay among all genes in the genome-wide context is essential to unravel the molecular mechanisms in the obesity etiology. However, as previously demonstrated there are differences between populations regarding to alleles frequencies. Belsky et al. [94] sought to develop a GRS for obesity using results obtained in 16 previously published GWAS in European descent samples. Analyzing 32 *locus* they found a significantly predictor of BMI and obesity among Europeans. However, the predictive effects for this GRS did not replicate among African Americans due particularly to the differences in risk-allele distributions.

Final Remarks

Obesity is a complex phenotype resulting from the interaction of several internal and external factors. Although most scientists and clinicians now acknowledge that genes contribute to obesity, at this point relatively little is known regarding the specific *loci* involved and the mechanism by which they lead to the expression of obesity. Like many other complex human traits, environmental factors also play a major role in the etiology of obesity. Evidence suggests that interactions between genetic and environmental factors may contribute to common obesity. The contribution of new techniques as whole exome and whole genome analysis could play an important role to identified new and rare polymorphisms associated with common obesity.

References

1. Haslam DW, James WPT (2005) Obesity. Lancet 366:1197–1209
2. Swinburn B, Sacks G, Hall KD, McPherson K, Finegood DT, Moodie ML et al (2011) The global obesity pandemic: shaped by global drivers and local environments. Lancet 378:804–814
3. Silventoinen K, Rokholm B, Kaprio J, Sørensen TIA (2010) The genetic and environmental influences on childhood obesity: a systematic review of twin and adoption studies. Int J Obes (Lond) 34:29–40
4. Sandholt CH, Hansen T, Pedersen O (2012) Beyond the fourth wave of genome-wide obesity association studies. Nutr Diabetes 2:e37
5. Mathes WF, Aylor DL, Miller DR, Churchill GA, Chesler EJ, de Villena FP-M et al (2011) Architecture of energy balance traits in emerging lines of the Collaborative Cross. Am J Physiol Endocrinol Metab 300:e1124–e1134
6. Finucane MM, Stevens G, Cowan MJ, Danaei G, Lin JK, Paciorek CJ et al (2011) National, regional, and global trends in body-mass index since 1980: systematic analysis of health examination surveys and epidemiological studies with 960 country-years and 9·1 million participants. Lancet 377:557–567
7. Xia Q, Grant SFA (2013) The genetics of human obesity. Ann N Y Acad Sci 1281:178–190
8. O'Rahilly S, Farooqi IS (2008) Human obesity: a heritable neurobehavioral disorder that is highly sensitive to environmental conditions. Diabetes 57:2905–2910
9. Stunkard AJ, Sørensen TI, Hanis C, Teasdale TW, Chakraborty R, Schull WJ et al (1986) An adoption study of human obesity. N Engl J Med 314:193–198
10. Stunkard AJ, Foch TT, Hrubec Z (1986) A twin study of human obesity. JAMA 256:51–54
11. Feinleib M, Garrison RJ, Fabsitz R, Christian JC, Hrubec Z, Borhani NO et al (1977) The NHLBI twin study of cardiovascular disease risk factors: methodology and summary of results. Am J Epidemiol 106:284–285

12. Whitaker RC, Wright JA, Pepe MS, Seidel KD, Dietz WH (1997) Predicting obesity in young adulthood from childhood and parental obesity. N Engl J Med 337:869–873
13. Danielzik S, Langnäse K, Mast M, Spethmann C, Müller MJ (2002) Impact of parental BMI on the manifestation of overweight 5–7 year old children. Eur J Nutr 41:132–138
14. Magnusson PKE, Rasmussen F (2002) Familial resemblance of body mass index and familial risk of high and low body mass index. A study of young men in Sweden. Int J Obes Relat Metab Disord 26:1225–1231
15. Mamun AA, O'Callaghan M, Callaway L, Williams G, Najman J, Lawlor DA (2009) Associations of gestational weight gain with offspring body mass index and blood pressure at 21 years of age: evidence from a birth cohort study. Circulation 119:1720–1727
16. Knowler WC, Pettitt DJ, Saad MF, Bennett PH (1990) Diabetes mellitus in the Pima Indians: incidence, risk factors and pathogenesis. Diabetes Metab Rev 6:1–27
17. Rankinen T, Zuberi A, Chagnon YC, Weisnagel SJ, Argyropoulos G, Walts B et al (2006) The human obesity gene map: the 2005 update. Obesity (Silver Spring) 14:529–644
18. Hinney A, Vogel CIG, Hebebrand J (2010) From monogenic to polygenic obesity: recent advances. Eur Child Adolesc Psychiatry 19:297–310
19. Razquin C, Marti A, Martinez JA (2011) Evidences on three relevant obesogenes: MC4R, FTO and PPARγ. Approaches for personalized nutrition. Mol Nutr Food Res 55:136–149
20. Boutin P, Dina C, Vasseur F, Dubois S, Corset L, Séron K et al (2003) GAD2 on chromosome 10p12 is a candidate gene for human obesity. PLoS Biol 1:E68
21. Meyre D, Bouatia-Naji N, Tounian A, Samson C, Lecoeur C, Vatin V et al (2005) Variants of ENPP1 are associated with childhood and adult obesity and increase the risk of glucose intolerance and type 2 diabetes. Nat Genet 37:863–867
22. Suviolahti E, Oksanen LJ, Ohman M, Cantor RM, Ridderstrale M, Tuomi T et al (2003) The SLC6A14 gene shows evidence of association with obesity. J Clin Invest 112:1762–1772
23. Hinney A, Hebebrand J (2008) Polygenic obesity in humans. Obes Facts 1:35–42
24. Day FR, Loos RJF (2011) Developments in obesity genetics in the era of genome-wide association studies. J Nutrigenet Nutrigenomics 4:222–238
25. Bell CG, Walley AJ, Froguel P (2005) The genetics of human obesity. Nat Rev Genet 6:221–234
26. Pérusse L, Rankinen T, Zuberi A, Chagnon YC, Weisnagel SJ, Argyropoulos G et al (2005) The human obesity gene map: the 2004 update. Obes Res 13:381–490
27. Loos RJF (2012) Genetic determinants of common obesity and their value in prediction. Best Pract Res Clin Endocrinol Metab 26:211–226
28. Mutch DM, Clément K (2006) Genetics of human obesity. Best Pract Res Clin Endocrinol Metab 20:647–664
29. Ramachandrappa S, Farooqi IS (2011) Review series Genetic approaches to understanding human obesity. 121. doi:10.1172/JCI46044.2080
30. Marian AJ (2012) Molecular genetic studies of complex phenotypes. Transl Res 159:64–79
31. Mardis ER (2013) Next-generation sequencing platforms. Annu Rev Anal Chem (Palo Alto Calif) 6:287–303
32. Saeed S, Bonnefond A, Manzoor J, Philippe J, Durand E, Arshad M et al (2014) Novel LEPR mutations in obese Pakistani children identified by PCR-based enrichment and next generation sequencing. Obesity (Silver Spring) 22:1112–1117
33. Sällman Almén M, Rask-Andersen M, Jacobsson JA, Ameur A, Kalnina I, Moschonis G et al (2013) Determination of the obesity-associated gene variants within the entire FTO gene by ultra-deep targeted sequencing in obese and lean children. Int J Obes (Lond) 37: 424–431
34. Herbert A, Gerry NP, McQueen MB, Heid IM, Pfeufer A, Illig T et al (2006) A common genetic variant is associated with adult and childhood obesity. Science 312:279–283
35. Frayling TM, Timpson NJ, Weedon MN, Zeggini E, Freathy RM, Lindgren CM et al (2007) A common variant in the FTO gene is associated with body mass index and predisposes to childhood and adult obesity. Science 316:889–894
36. Albuquerque D, Nóbrega C, Manco L (2013) Association of FTO polymorphisms with obesity and obesity-related outcomes in Portuguese children. PLoS One 8:e54370

37. Rodríguez-López R, González-Carpio M, Serrano MV, Torres G, García de Cáceres MT, Herrera T et al (2010) Association of FTO gene polymorphisms and morbid obesity in the population of Extremadura (Spain). Endocrinol Nutr 57:203–209
38. Chang Y-C, Liu P-H, Lee W-J, Chang T-J, Jiang Y-D, Li H-Y et al (2008) Common variation in the fat mass and obesity-associated (FTO) gene confers risk of obesity and modulates BMI in the Chinese population. Diabetes 57:2245–2252
39. Hotta K, Nakata Y, Matsuo T, Kamohara S, Kotani K, Komatsu R et al (2008) Variations in the FTO gene are associated with severe obesity in the Japanese. J Hum Genet 53:546–553
40. Fang H, Li Y, Du S, Hu X, Zhang Q, Liu A et al (2010) Variant rs9939609 in the FTO gene is associated with body mass index among Chinese children. BMC Med Genet 11:136
41. Mačeková S, Bernasovský I, Gabriková D, Bôžiková A, Bernasovská J, Boroňová I et al (2012) Association of the FTO rs9939609 polymorphism with obesity in Roma/Gypsy population. Am J Phys Anthropol 147:30–34
42. Grant SFA, Li M, Bradfield JP, Kim CE, Annaiah K, Santa E et al (2008) Association analysis of the FTO gene with obesity in children of Caucasian and African ancestry reveals a common tagging SNP. PLoS One 3:e1746
43. Song Y, You N-C, Hsu Y-H, Howard BV, Langer RD, Manson JE et al (2008) FTO polymorphisms are associated with obesity but not diabetes risk in postmenopausal women. Obesity (Silver Spring) 16:2472–2480
44. Deliard S, Panossian S, Mentch FD, Kim CE, Hou C, Frackelton EC et al (2013) The missense variation landscape of FTO, MC4R, and TMEM18 in obese children of African Ancestry. Obesity (Silver Spring) 21:159–163
45. Dina C, Meyre D, Gallina S, Durand E, Körner A, Jacobson P et al (2007) Variation in FTO contributes to childhood obesity and severe adult obesity. Nat Genet 39:724–726
46. Scuteri A, Sanna S, Chen W-M, Uda M, Albai G, Strait J et al (2007) Genome-wide association scan shows genetic variants in the FTO gene are associated with obesity-related traits. PLoS Genet 3:e115
47. Heard-Costa NL, Zillikens MC, Monda KL, Johansson A, Harris TB, Fu M et al (2009) NRXN3 is a novel locus for waist circumference: a genome-wide association study from the CHARGE Consortium. PLoS Genet 5:e1000539
48. Lindgren CM, Heid IM, Randall JC, Lamina C, Steinthorsdottir V, Qi L et al (2009) Genome-wide association scan meta-analysis identifies three Loci influencing adiposity and fat distribution. PLoS Genet 5:e1000508
49. Kilpeläinen TO, Qi L, Brage S, Sharp SJ, Sonestedt E, Demerath E et al (2011) Physical activity attenuates the influence of FTO variants on obesity risk: a meta-analysis of 218,166 adults and 19,268 children. PLoS Med 8:e1001116
50. Tung Y-CL, Yeo GSH (2011) From GWAS to biology: lessons from FTO. Ann N Y Acad Sci 1220:162–171
51. Peters T, Ausmeier K, Dildrop R, Rüther U (2002) The mouse Fused toes (Ft) mutation is the result of a 1.6-Mb deletion including the entire Iroquois B gene cluster. Mamm Genome 13:186–188
52. Fischer J, Koch L, Emmerling C, Vierkotten J, Peters T, Brüning JC et al (2009) Inactivation of the Fto gene protects from obesity. Nature 458:894–898
53. Church C, Lee S, Bagg EAL, McTaggart JS, Deacon R, Gerken T et al (2009) A mouse model for the metabolic effects of the human fat mass and obesity associated FTO gene. PLoS Genet 5:e1000599
54. Church C, Moir L, McMurray F, Girard C, Banks GT, Teboul L et al (2010) Overexpression of Fto leads to increased food intake and results in obesity. Nat Genet 42:1086–1092
55. Gerken T, Girard CA, Tung Y-CL, Webby CJ, Saudek V, Hewitson KS et al (2007) The obesity-associated FTO gene encodes a 2-oxoglutarate-dependent nucleic acid demethylase. Science 318:1469–1472
56. van den Born E, Omelchenko MV, Bekkelund A, Leihne V, Koonin EV, Dolja VV, Falnes PØ (2008) Viral AlkB proteins repair RNA damage by oxidative demethylation. Nucleic Acids Res 36(17):5451–5461

57. Jia G, Fu Y, Zhao X, Dai Q, Zheng G, Yang Y et al (2011) N6-methyladenosine in nuclear RNA is a major substrate of the obesity-associated FTO. Nat Chem Biol 7:885–887
58. Speakman JR, Rance KA, Johnstone AM (2008) Polymorphisms of the FTO gene are associated with variation in energy intake, but not energy expenditure. Obesity (Silver Spring) 16:1961–1965
59. Jacobsson J a, Schiöth HB, Fredriksson R (2012) The impact of intronic single nucleotide polymorphisms and ethnic diversity for studies on the obesity gene FTO. Obes Rev 13:1096–1109
60. Han Z, Niu T, Chang J, Lei X, Zhao M, Wang Q et al (2010) Crystal structure of the FTO protein reveals basis for its substrate specificity. Nature 464:1205–1209
61. Fawcett KA, Barroso I (2010) The genetics of obesity: FTO leads the way. Trends Genet 26:266–274
62. Loos RJF, Lindgren CM, Li S, Wheeler E, Zhao JH, Prokopenko I et al (2008) Common variants near MC4R are associated with fat mass, weight and risk of obesity. Nat Genet 40:768–775
63. Farooqi IS, O'Rahilly S (2005) Monogenic obesity in humans. Annu Rev Med 56:443–458
64. Chambers JC, Elliott P, Zabaneh D, Zhang W, Li Y, Froguel P et al (2008) Common genetic variation near MC4R is associated with waist circumference and insulin resistance. Nat Genet 40:716–718
65. Thorleifsson G, Walters GB, Gudbjartsson DF, Steinthorsdottir V, Sulem P, Helgadottir A et al (2009) Genome-wide association yields new sequence variants at seven loci that associate with measures of obesity. Nat Genet 41:18–24
66. Xi B, Chandak GR, Shen Y, Wang Q, Zhou D (2012) Association between common polymorphism near the MC4R gene and obesity risk: a systematic review and meta-analysis. PLoS One 7:e45731
67. Willer CJ, Speliotes EK, Loos RJF, Li S, Lindgren CM, Heid IM et al (2009) Six new loci associated with body mass index highlight a neuronal influence on body weight regulation. Nat Genet 41:25–34
68. Speliotes EK, Willer CJ, Berndt SI, Monda KL, Thorleifsson G, Jackson AU et al (2010) Association analyses of 249,796 individuals reveal 18 new loci associated with body mass index. Nat Genet 42:937–948
69. Berndt SI, Gustafsson S, Mägi R, Ganna A, Wheeler E, Feitosa MF et al (2013) Genome-wide meta-analysis identifies 11 new loci for anthropometric traits and provides insights into genetic architecture. Nat Genet 45:501–512
70. Heid IM, Jackson AU, Randall JC, Winkler TW, Qi L, Steinthorsdottir V et al (2010) Meta-analysis identifies 13 new loci associated with waist-hip ratio and reveals sexual dimorphism in the genetic basis of fat distribution. Nat Genet 42:949–960
71. Albuquerque D, Nóbrega C, Manco L (2013) The lactase persistence − 13910C>T polymorphism shows indication of association with abdominal obesity among Portuguese children. Acta Paediatr 102(4):153–157
72. Kettunen J, Silander K, Saarela O, Amin N, Müller M, Timpson N et al (2010) European lactase persistence genotype shows evidence of association with increase in body mass index. Hum Mol Genet 19:1129–1136
73. Corella D, Arregui M, Coltell O, Portolés O, Guillem-Sáiz P, Carrasco P et al (2011) Association of the LCT-13910C>T polymorphism with obesity and its modulation by dairy products in a Mediterranean population. Obesity (Silver Spring) 19:1707–1714
74. Almon R, Álvarez-León EE, Serra-Majem L (2012) Association of the European lactase persistence variant (LCT-13910 C>T polymorphism) with obesity in the Canary Islands. PLoS One 7:e43978
75. Kilpeläinen TO, Zillikens MC, Stančákova A, Finucane FM, Ried JS, Langenberg C et al (2011) Genetic variation near IRS1 associates with reduced adiposity and an impaired metabolic profile. Nat Genet 43:753–760

76. Yang J, Manolio T a, Pasquale LR, Boerwinkle E, Caporaso N, Cunningham JM et al (2011) Genome partitioning of genetic variation for complex traits using common SNPs. Nat Genet 43:519–525
77. Llewellyn CH, Trzaskowski M, Plomin R, Wardle J (2013) Finding the missing heritability in pediatric obesity: the contribution of genome-wide complex trait analysis. Int J Obes (Lond) 37:1506–1509
78. Smemo S, Tena JJ, Kim K-H, Gamazon ER, Sakabe NJ, Gómez-Marín C et al (2014) Obesity-associated variants within FTO form long-range functional connections with IRX3. Nature 507:371–375
79. Albuquerque D, Nóbrega C, Rodríguez-López R, Manco L (2014) Association study of common polymorphisms in MSRA, TFAP2B, MC4R, NRXN3, PPARGC1A, TMEM18, SEC16B, HOXB5 and OLFM4 genes with obesity-related traits among Portuguese children. J Hum Genet 59(6):307–313
80. Zhao J, Bradfield JP, Zhang H, Sleiman PM, Kim CE, Glessner JT et al (2011) Role of BMI-associated loci identified in GWAS meta-analyses in the context of common childhood obesity in European Americans. Obesity (Silver Spring) 19:2436–2439
81. León-Mimila P, Villamil-Ramírez H, Villalobos-Comparán M, Villarreal-Molina T, Romero-Hidalgo S, López-Contreras B et al (2013) Contribution of common genetic variants to obesity and obesity-related traits in Mexican children and adults. PLoS One 8:e70640
82. Zhao J, Grant SF (2011) Genetics of childhood obesity. J Obes 2011:845148.
83. Mitchell JA, Hakonarson H, Rebbeck TR, Grant SFA (2013) Obesity-susceptibility loci and the tails of the pediatric BMI distribution. Obesity (Silver Spring) 21:1256–1260
84. Bradfield JP, Taal HR, Timpson NJ, Scherag A, Lecoeur C, Warrington NM et al (2012) A genome-wide association meta-analysis identifies new childhood obesity loci. Nat Genet 44:526–531
85. Ntalla I, Panoutsopoulou K, Vlachou P, Southam L, William Rayner N, Zeggini E et al (2013) Replication of established common genetic variants for adult BMI and childhood obesity in Greek adolescents: the TEENAGE study. Ann Hum Genet 77:268–274
86. Slatkin M (2008) Linkage disequilibrium-understanding the evolutionary past and mapping the medical future. Nat Rev Genet 9:477–485
87. Hennig BJ, Fulford AJ, Sirugo G, Rayco-Solon P, Hattersley AT, Frayling TM et al (2009) FTO gene variation and measures of body mass in an African population. BMC Med Genet 10:21
88. Ohashi J, Naka I, Kimura R, Natsuhara K, Yamauchi T, Furusawa T et al (2007) FTO polymorphisms in oceanic populations. J Hum Genet 52:1031–1035
89. Hassanein MT, Lyon HN, Nguyen TT, Akylbekova EL, Waters K, Lettre G et al (2010) Fine mapping of the association with obesity at the FTO locus in African-derived populations. Hum Mol Genet 19:2907–2916
90. Wen W, Cho Y-S, Zheng W, Dorajoo R, Kato N, Qi L et al (2012) Meta-analysis identifies common variants associated with body mass index in east Asians. Nat Genet 44:307–311
91. Monda KL, Chen GK, Taylor KC, Palmer C, Edwards TL, Lange LA et al (2013) A meta-analysis identifies new loci associated with body mass index in individuals of African ancestry. Nat Genet 45:690–696
92. Tan L-J, Zhu H, He H, Wu K-H, Li J, Chen X-D et al (2014) Replication of 6 obesity genes in a meta-analysis of genome-wide association studies from diverse ancestries. PLoS One 9:e96149
93. Horne BD, Anderson JL, Carlquist JF, Muhlestein JB, Renlund DG, Bair TL et al (2005) Generating genetic risk scores from intermediate phenotypes for use in association studies of clinically significant endpoints. Ann Hum Genet 69:176–186
94. Belsky DW, Moffitt TE, Sugden K, Williams B, Houts R, McCarthy J et al (2013) Development and evaluation of a genetic risk score for obesity. Biodemography Soc Biol 59:85–100

Chapter 5
The Role of the GWAS Identified *FTO* Locus in Regulating Body Size and Composition

Giles S. H. Yeo

Obesity is a Problem… Who Can We Blame?

Since time immemorial, the control of food intake and body-weight has been thought to be simply an issue of self-control and will-power. Gluttony is, after all, one of the 'seven deadly sins'. So as obesity has become an increasing public health problem, reaching epidemic proportions in most developed and emerging economies, society in turn blames those overweight and obese for a lack of moral fortitude. 'The obese only have themselves to blame, all they have to do is to eat less….' so the argument goes.

From a thermodynamic standpoint, this view is of course quite accurate. Body-weight is clearly a balance between energy intake and energy expenditure. Thus, the only way to gain weight is to eat more than you burn, and the only way to lose weight is to eat less than you burn. There, in one succinct statement, is the cause and cure of the obesity epidemic. However, this sage piece of advice that your grandmother could have given you, is clearly not working. A far more complex and interesting question to ask is WHY some people eat more than others. Few would dispute that the obesity epidemic has been driven by lifestyle and environmental changes. However individuals respond differently to these 'obesigenic' environmental changes and this variation in response has a strong genetic element underlying physiological variations. Indeed studies of BMI (Body Mass Index; weight in kg/height in m^2; a correlate of body fat mass) correlations of monozygotic, dizygotic, biological and adopted siblings reveal heritability of fat mass to be between 40–70 % [1, 2]. Consequently, genetic approaches offer an effective tool for characterising the molecular

G. S. H. Yeo (✉)
MRC Metabolic Diseases Unit, Metabolic Research Laboratories,
University of Cambridge, Level 4, Wellcome Trust-MRC Institute of Metabolic Science,
Box 289, Addenbrooke's Hospital, CB2 0QQ Cambridge, UK
e-mail: gshy2@cam.ac.uk

C. Nóbrega, R. Rodriguez-López (eds.), *Molecular Mechanisms Underpinning the Development of Obesity,* DOI 10.1007/978-3-319-12766-8_5

and physiological mechanisms of food intake and body weight control, and allow us to understand how these may become defective in the obese state.

Genetics as a Tool

The study of extreme phenotypes in both mice and humans, have identified a number of genes that when mutated, cause severe obesity. Although relatively rare, these monogenic disorders indicate a fundamental failure of the mechanisms of energy homeostasis, and together with genetically modified mouse models, have illuminated the critical role that these molecules play in the physiological control of food intake and bodyweight [3]. However, the genetic determinants of body weight in the general population are believed largely to be polygenic and only recently has the spate of genome-wide association studies (GWAS) begun to reveal some of the genetic architecture underlying common obesity.

Identification of the Fat Mass and Obesity Associated (*FTO*) Locus

In 2007, a GWAS identified a cluster of single nucleotide polymorphisms (SNPs) on chromosome 16q12 to be associated with type 2 diabetes (T2D) in Europeans [4]. However, after adjusting for BMI, the association with T2D was completely abolished, suggesting that the association of these SNPs with T2D was mediated through their effect on BMI. The cluster of SNPs are located within intron 1 of *Fat Mass and Obesity Associated (FTO)*, with the surrounding region including the genes *IRX3, 5, 6, FTM* and *FTL* (more on these later; Fig. 5.1). The study initially compared 1900 cases and 3000 controls, while follow-up analyses in 39,000 individuals confirmed the association with BMI and obesity risk [4]. Eight weeks after the first discovery, a GWAS for BMI in 4700 Sardinians observed highly significant associations for SNPs from the same intronic cluster in *FTO*, which was subsequently replicated in 2300 European and Hispanic Americans [5]. A third study, published at around the same time as the two GWAS, identified the same *FTO* locus serendipitously, while testing for population stratification in their case-control obesity data [6]. Together, these studies firmly established *FTO* as the first gene with common variants, albeit non-coding, that affect obesity susceptibility in the general population.

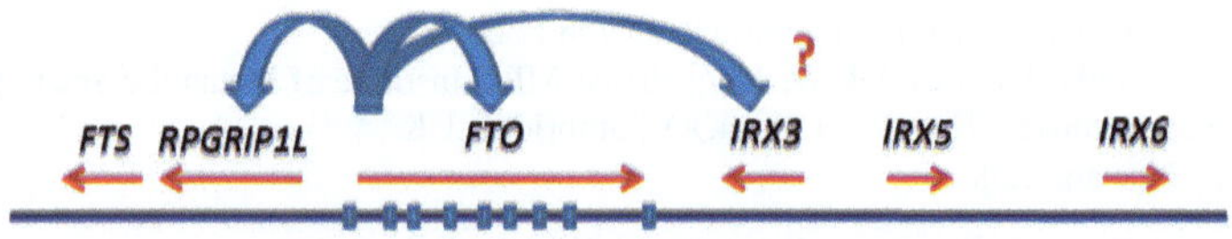

Fig. 5.1 Schematic representation of chromosome 16q22

The *FTO* Locus—A Cluster of BMI-Associated SNPs in *FTO's* First Intron

The two GWAS that first reported on *FTO* as an obesity susceptibility gene each identified a different SNP in the first intron as the most significantly associated with BMI; i.e. rs9939609 [4], rs9930506 [5]. Subsequent GWAS studies for obesity-related traits in European ancestry populations all confirmed the *FTO* locus, but reported a number of other *FTO* SNPs, located in the same chromosomal region [7–15]. All these GWAS-identified *FTO* SNPs are part of the same cluster of highly correlated SNPs and, as a consequence, they are all highly significantly associated with BMI and other obesity related traits. In European ancestry populations, this SNP cluster stretches across ~46 kilo base-pairs (kb) in *FTO's* first intron that likely harbors the causal variant(s).

Replication Across Multiple Ethnicities and Age Groups

Soon after the discovery of *FTO* in European ancestry populations, many replication efforts examined its effects in non-European ancestry populations, showing convincing support for the generalizability of *FTO* as an obesity-susceptibility locus across most ancestries studied so far. The most consistent replications have been observed for Asian ancestry populations. Three large-scale GWAS in East Asians identified *FTO* SNPs as the most significantly associated with BMI [16–18]. In addition, targeted efforts consistently confirmed association with obesity-related traits in multiple other Asian populations, including Vietnamese, Filipino, Malays and Indian Asians (reviewed by [19]). SNPs in *FTO* also show association with obesity-related traits in Hispanic/Latino populations and in Pima Indians [19].

Results of targeted replication of *FTO* in Africans and African Americans have been inconsistent [20–22], which may be due to the substantial differences in correlation structure between *FTO* SNPs in African compared to European/Asian ancestry populations. As such, an *FTO* SNP that is part of the larger 'European/Asian' cluster, but that does not overlap with the 'African' cluster, will likely not show association with obesity related traits. However, in the recent large-scale GWAS in African ancestry populations, SNPs in *FTO* were among the most significantly associated with BMI, firmly establishing *FTO* as an obesity-susceptibility locus also in this ancestry [23]. While *FTO's* effect on BMI was similar to that observed in European ancestry populations, the minor allele frequency was much lower (12 %), such that only 0.10 % of the variation of BMI in African ancestry populations was explained [23].

Although *FTO* was first discovered as an obesity-susceptibility locus in adults [4–6], its associations with obesity related traits were promptly confirmed in studies of children and adolescents [4, 8, 10, 11]. While SNPs in *FTO* do not influence birth weight [24–26], longitudinal studies have shown that they already affect body weight during early childhood, as early as age 3 years, after which *FTO*'s effect increases to reach its largest impact at around young adulthood [27–29].

Taken together, SNPs in *FTO* show association with obesity-related traits across many ancestries and age groups. Noteworthy is that the genotype frequency distribution of the BMI-associated *FTO* SNPs differs substantially across ancestries, with the highest prevalence of minor (risk) allele carriers observed in European ancestry populations, and substantially fewer in Asian and African ancestry populations [19].

Effect Size

Since the discovery of *FTO* in 2007, at least many additional obesity-susceptibility loci have been identified through large-scale GWAS efforts [12]. However, the *FTO* locus stands out, as it has by far the largest effect size, is very common, and has the largest explained variance among individuals of European ancestry. More specifically, each additional minor (risk) allele is associated with a 0.39 kg/m^2 higher BMI (equivalent to 1130 g for a person of 1.70 m tall) and a 1.20 fold increased risk of obesity [12]. Approximately 43 % of the population carries one risk allele and 20 % carries two risk alleles, with small variations in genotype frequencies within European ancestry populations. Of all BMI-associated loci identified thus far, the *FTO* locus explains the most of the inter-individual variation in BMI, yet only a mere 0.34 % [12]. As a consequence, the ability to predict a person's risk of obesity based on their *FTO* genotype is poor and only slightly better than tossing a coin [19].

The Role of Lifestyle Factors in the Association Between *FTO* and Obesity-Susceptibility

To gain insight in the potential mechanisms through which variation in the *FTO* locus leads to increased risk of obesity, studies then considered whether *FTO* SNPs associate with the two major mediators of body weight regulation, food intake and physical activity.

The evidence supporting a role for *FTO* in the regulation of food intake is compelling. The BMI-increasing allele of *FTO* SNPs has been found to be associated with increased energy intake [30–32], increased intake of dietary fat [32, 33] or protein [34], increased appetite and reduced satiety [35, 36], and loss of control over eating [37]. A recent GWAS of macronutrient intake in more than 70,000 individuals identified the BMI-increasing allele of *FTO* SNPs to be highly significantly associated with increased protein intake [38].

However, studies have consistently shown that *FTO* SNPs are not associated with physical activity levels [31, 39–41], which has been convincingly confirmed in a large-scale meta-analysis of published and unpublished data of more than 200,000 adults and 20,000 children [42]. Although physical activity does not seem to mediate the association between *FTO* and obesity-susceptibility, this meta-analysis showed that the effect of *FTO* on BMI and obesity risk is approximately 30% smaller in physically active than in sedentary individuals, at least in adults [42]. This is consistent with observations that between-person variability in obesity can also partly be explained by the genotype at the *FTO* locus [43]. In other words, although carriers of the *FTO* risk allele have an increased mean BMI, there is also increased variability of effect size around the mean. This is likely influenced by the impact of exercise and other environmental factors, highlighting the importance of physical activity in influencing the 'genetic burden' of the *FTO* risk alleles.

Thus, while studying lifestyle factors such as physical activity and food intake is challenging because of the inaccuracy of their measurement, there is growing evidence that physical activity attenuates the association between *FTO* and obesity susceptibility, whereas food intake might be mediating this association. It remains to be confirmed which components of food intake are predominantly targeted by *FTO*.

Is it *FTO*?

GWAS are by nature gene agnostic, and SNPs reaching the appropriate statistical threshold for a given phenotype can appear anywhere in the genome, within, near or far away from any coding sequence. The current assumption that the closest coding region, which is sometimes hundreds of kilobases away, is the likely candidate is perhaps a reasonable first guess, but not necessarily true [44]. Thus a major challenge in the field has been to translate these statistical hits into real biological insight. As mentioned at the outset, in addition to *FTO,* the surrounding locus on chromosome 16q22 also encompasses the Iroquois B cluster of *IRX*3, 5, and 6, as well as *FTM* and *FTL* (Fig. 5.1). The key question is which of these genes are responsible for the association with obesity, and what is the underlying mechanism?

The two closest genes in this instance were; with the obesity risk alleles actually located in its first intron, *FTO*; and, with the risk SNPs in close proximity to its transcriptional start site adjacent to and coded for on the opposite DNA strand, *RPGRIP1L*(human ortholog of mouse *Ftm*) [45]. Between these, there were two main reasons why the focus of study over the past few years has been with *FTO* and not *RPGRIP1L* Firstly, while *FTO* was found to be nutritionally regulated within the hypothalamus [46, 47], this was not true for RPGRIP1L which is known to localize in the primary cilia and centrosomes of ciliated cells. Secondly, human defects in *RPGRIP1L* exist and cause Joubert syndrome type 7 (JBTS), which presents clinically with cerebellar and brainstem malformation and renal failure [48]. The patients do not present with any obvious body weight-related phenotypes, with the caveat that any potential 'lean' phenotype is difficult to ascertain in a healthy individual, let alone someone who is severely ill. Deletion of the mouse ortholog *Rpgrip1l* (*Ftm*) recapitulates the cerebral, renal and hepatic defects seen in Joubert's patients [48].

FTO is a Nucleic Acid Demethylase

In the same year of its identification as a candidate gene for obesity, *FTO* was predicted, using bioinformatics tools, to be a 2-oxyglutarate (2-OG) Fe(II) dependent demethylase, closely related to the bacterial DNA demethylase AlkB and the mammalian AlkB homologues 1 & 2 (ABH1 and ABH2) [46]. In vitro, recombinant *FTO* is able to catalyze the Fe(II)- and 2OG-dependent demethylation of 3-methylthymine in single-stranded DNA, as well as 3-methyluracil (3meU) [46, 49] and 6-methyl adenosine (6meA) [50] in single-stranded RNA, suggesting a potential role for *FTO* in nucleic acid repair or modification (Fig. 5.2). The crystal structure of *FTO* is available and shows an N-terminal catalytic domain and a C-terminal domain of unknown function [51]. The catalytic pocket contains five amino acid residues invariant in all members of this enzyme superfamily; two residues, a histidine (H) and an aspartic acid (D), required for binding Fe(II); and three residues, an H and two arginines (R) separated by six amino acids, required for 2OG binding [51, 52]. The specificity for single stranded nucleic acids is provided by an L1 loop, not present in other members of the AlkB family, which sterically hinders double stranded nucleic acids from entering the catalytic pocket [51].

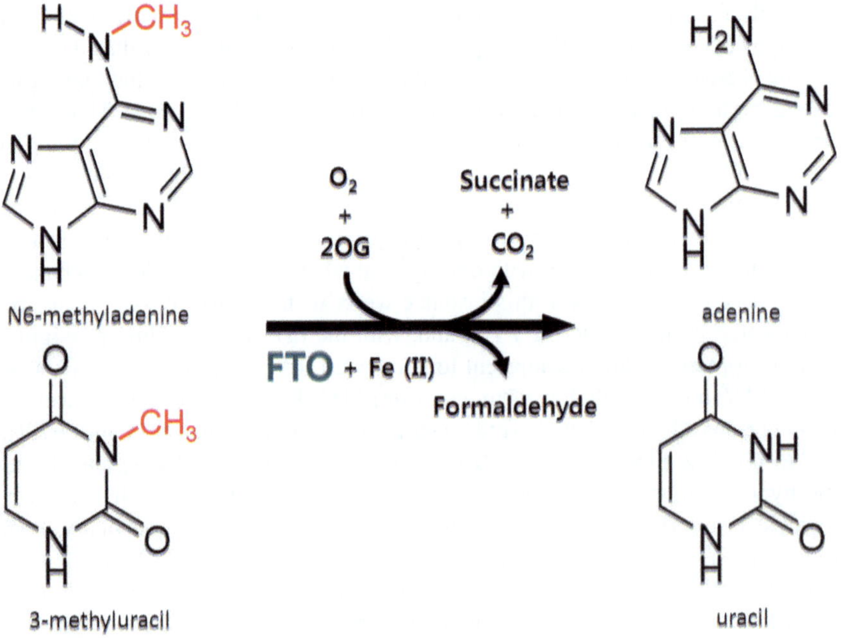

Fig. 5.2 Biochemical function of *FTO*. *FTO* is able to catalyze the Fe(II)- and 2OG-dependent demethylation of N6-methyladenine and 3-methyluracil with concomitant production of succinate, formaldehyde, and carbon dioxide. N6-methyladenine is found in mRNA while 3-methyluracil is found largely in ribosomal RNA

6meA, the most common modified nucleoside found in mRNA [53], is demethylated by *FTO* with 50-fold greater affinity than 3meU [50], which is found largely in ribosomal RNA [54]. However, with vast majority of total RNA composed of rRNA, there are in absolute terms actually a 100 fold more 3meU than 6meA in any given cell. The question of whether one or both of these modified bases are the endogenous substrate/s for *FTO* is still to be determined.

FTO Deficiency

There are notable examples where common variants close to a particular gene, such as *POMC* [55, 56], *PCSK1* [57, 58], *MC4R* [59, 60] and *BDNF* [13, 61] are associated with an increased risk of common phenotypes, such as susceptibility to obesity, whereas rare loss-of-function mutations in these same genes result in Mendelian forms of severe early onset obesity. Thus it was reasonable to consider what would happen in the case of *FTO* deficiency.

Fto was originally identified as one of six contiguous genes, together with *Ftm*, *Ftl* and *Irx*3, 5, and 6, in a 1.6 Mb deletion causing the 'fused-toe' phenotype [62]. Homozygotes are embryonically lethal, while heterozygous fused toes mutants display severe developmental defects including left-right asymmetry [63], defects in hypothalamic development [63, 64], the eponymous fused digits and hyperplasia of the thymus. In contrast, mice with a specific targeted deletion of *Fto* did not display these severe developmental abnormalities. Instead, *Fto* homozygous null mice exhibited a phenotype of postnatal growth retardation, decreased fat and lean body mass, and elevated food intake when corrected for lean body mass [65], accompanied by a significant level of post-natal lethality, with only 50% of homozygous pups surviving till weaning [65, 66].

In humans, a loss-of-function mutation in *FTO* leads to an even more complex phenotype of postnatal growth retardation, microcephaly, severe psychomotor delay, functional brain deficits, and characteristic facial dysmorphism [67]. Structural brain malformations, cardiac defects, genital anomalies, and cleft palate were also observed in some of the patients. The R316Q mutation disrupts one of the obligate arginines necessary for 2-OG binding and results in the loss of *FTO*'s demethylase activity. The importance of *FTO*'s ability to demethylate is underlined not only by the severe phenotype detailed above, but also by the tragic fact that none of the affected individuals survived past the age of 30 months [67]. So in both humans and mice, a fully functional *FTO* certainly appears to be critical for normal physiology.

Perturbation of *FTO* Expression Points to Role in Energy Homeostasis

Even despite the severe phenotype seen in *FTO* deficiency however, a compelling body of evidence has emerged to support a role for *FTO* in the control of energy homeostasis. As briefly touched on above, initial observations reported that

Fto-/- mice have an apparent 'hyperphagia', and there is the fact that *Fto+/-* mice are resistant to the effects of a high-fat diet [65]. In contrast, Church and colleagues have generated a 'knock-in' mouse model carrying one or two additional copies of *Fto*, and show that ubiquitous overexpression of *Fto* leads to a dose-dependent increase in body and fat mass [68]. That said, although the increase in weight with the overexpression of *Fto* seems consistent with the 'lean' phenotype of *FTO* deficiency, the increase in food intake seen in these mice is not. At 8 weeks, the mice overexpressing *Fto* do have reduced fasting leptin levels [68], which is odd given that obese mice with increased fat mass are normally expected to have increased leptin levels. One could speculate that the hyperphagic phenotype seen in this model is driven by the hypo-leptinaemia.

FTO is expressed ubiquitously in human and animal tissues [46], which is consistent with multiple organ systems being affected in *FTO* deficiency. Its highest expression however, is seen in the brain, including the hypothalamus [46], where control of food intake is centred [3]. Within the arcuate nucleus of the hypothalamus (ARC), *Fto* is bi-directionally regulated as a function of nutritional status; decreasing following a 48 h fast [46] and increasing after 10 weeks of exposure to a high fat diet, while modulating *Fto* levels specifically in the ARC can influence food intake [47].

Thus, although there are some inconsistencies between the different mouse models to be worked out, the weight of evidence supports the notion that FTO itself can influence energy homeostasis by having direct effect on food intake. These findings are in keeping with the association between the *FTO* risk alleles and increase in food intake. Central nervous system (CNS) specific *FTO* deleted mice have also been generated [69]. Surprisingly, these brain-specific *Fto* deficient mice recapitulate much of the phenotype of the whole-body knock-outs, suggesting that an important proportion of FTO's function, particularly its link to the regulation of energy homeostasis is mediated in the brain.

FTO as a Nutrient Sensor?

Given the evidence above that *FTO* is nutritionally regulated and can influence food intake, could FTO be acting as a nutrient sensor? Initial attention focused on 2-OG, a key intermediate in the citric-acid cycle, and a co-substrate of FTO. It was plausible that FTO could function as a sensor for intracellular concentrations of this metabolite and thus cellular metabolism. However, since typical intracellular concentrations of 2-OG are more than 10-fold higher than its calculated Km of 2.88 μM, it is unlikely that FTO's physiological role is to sense 2-OG, even if it is required as a co-factor for *FTO* activity [70].

FTO mRNA and protein levels however, are dramatically down-regulated by total amino-acid deprivation mouse and human cell-lines [71]. Strikingly, this regulation was seen only with essential amino-acids, suggesting that FTO might play a role in the sensing of essential amino-acid availability [71]. Could the regulation of

FTO expression by amino acids then be linked to the growth retardation phenotype seen in *FTO* deficiency?

FTO Links Amino Acid Availability and mTORC1 Signaling to Regulate Growth and Translation

In a recent study, it was shown that mouse embryonic fibroblasts (MEFs) derived from *Fto*$^{-/-}$ mice exhibit slower rates of growth and have reduced mRNA translation when compared to WT MEFs [72]. This seems to occur, at least in part, by maintenance of levels of Aminoacyl-tRNA synthetases (AARSs), as part of a large multimer complex known as the Multi-Synthetase Complex (MSC) [73], which tether free amino-acids to their cognate tRNAs and are one of the key modulators of translation. Consistent with the reduced rates of translation, *Fto*$^{-/-}$ *MEFs* have reduced protein levels of MSC components. The defects in mRNA translation and reduced levels of MSC components in *Fto*$^{-/-}$ MEFs are rescued by re-expressing FTO in these cells, implicating a role for FTO in regulating translation rates through maintenance of MSC protein levels within the cell [72].

In addition, cells lacking FTO display decreased activation of the mTORC1 pathway and increased autophagy, all of which makes mechanistic sense in explaining the growth retardation phenotype seen in *Fto-/-* mice [65] and in humans homozygous for loss-of-function *FTO* mutations [67]. The dramatic regulation of *FTO* by AAs [71] seems to be necessary for the cellular response to changing AA levels, as expression of exogenous *FTO* in cells renders them insensitive to AA deprivation by preventing the expected reduction in mTORC1 signaling [72].

Thus, a cell without *FTO* is one that thinks it is starving of amino acids, reducing mTORC1 signaling and increasing autophagy in an effort to ensure cellular survival by maintaining cellular energy levels (Fig. 5.3). The consequence of this in a whole organism is clearly illustrated in another recent study, when a targeted deletion of *Fto* in adult mice results in a loss of body weight, with all of the change in weight down to a dramatic loss in lean mass [66]. As skeletal muscle is the largest depot of protein in the body, it appears to be the most sensitive to the sudden removal of *Fto* and subsequent increase in autophagy. In contrast, when *Fto* is postnatally deleted specifically from the mediobasal hypothalamus, lean mass is unaffected, and the resulting weight loss, stemming from a change in food intake, is subtle [66].

FTO and the 6meA Methylome

Critically, *FTO*'s role in linking AA availability and mTORC1 signaling with appropriates rates of growth and translation is dependent upon its ability to demethylate [72]. The question of how and why this occurs remains to be answered, and is now an area of intense activity. One possibility could involve *FTO*'s ability to demethylate 6meA.

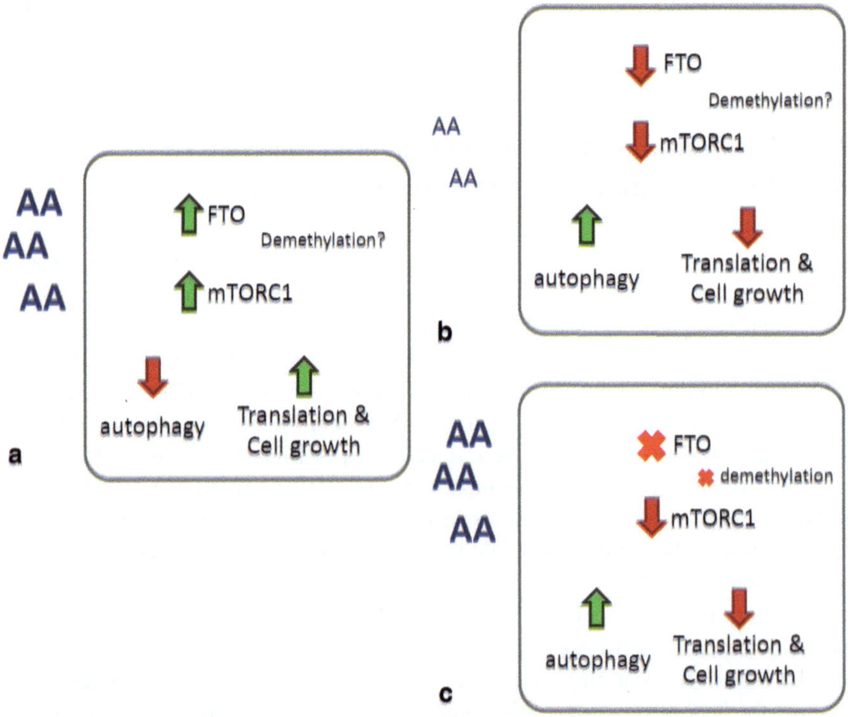

Fig. 5.3 Model of FTO's role in amino-acid sensing. **a** Under conditions of sufficient amino-acid levels, FTO levels and consequently mTORC1 activity are maintained, leading to appropriate translation and cell growth and inhibition of autophagy. **b** In conditions of amino-acid deprivation, FTO levels drop, leading to decreased mTORC1 activity, mRNA translation and cell growth, whereas autophagy is increased. This mechanism ensures cellular survival by maintaining cellular energy levels. **c** A cell with defective FTO (*red* X) interprets this as amino-acid starvation, thereby reducing mTORC1 signalling and increasing autophagy. Abbreviations: *AA* amino acid, FTO fat mass and obesity associated, *mTORC1* mammalian target of rapamycin complex 1

Using antibodies against 6meA to immunoprecipitate human and mouse transcripts that carry the modification and coupling this to RNA-seq, two recently published studies have mapped the presence of 6meA in a transcriptome-wide manner [74, 75]. 6meAs, as it turns out, are common, enriched near stop-codons, highly conserved between mice and man, and are dynamically, developmentally and tissue specifically regulated. These studies have far reaching implications, as the decoration of 6meA at appropriate mRNA sites appears to play a fundamental regulatory role in gene expression, in addition to exerting varying effects on mRNA splicing and transport. To date, methyltransferase like 3 (METTL3) is the only enzyme identified to catalyse the conversion of adenosine to 6meA [76], while two enzymes are known to catalyse the removal of this methyl group; one is ALKBH5 [77] and the other is FTO [50]. In fact, Meyer and colleagues show that transient overexpression of *FTO* in *HEK293* cells decreases the total amount of 6meA found in the transcriptome [75].

A recent paper has demonstrated that FTO does not globally target all 6meA-modified mRNAs but instead demethylates specific mRNA subsets [78]. They show for instance, that the midbrain and striatum of *Fto-/-* mice show increased methylation in mRNAs from genes encoding components of the dopamine signalling pathway, resulting in reduced dopaminergic signalling tone [78]. Outside of the brain, *FTO* risk alleles have also been reported to influence the methylation status of ghrelin mRNA. By doing so, Karra and colleagues have postulated that FTO may affect levels of circulating ghrelin. Further studies will be required to determine the extent to which changes in dopamine and/or ghrelin signalling could be effectors of *FTO's* association with increased BMI [79].

.... and Then Entering Stage Right, IRX3

Most recently, and perhaps heralding a sea-change in the way we begin ascribing the names of genes to GWAS identified disease associated SNPs, Smemo and colleagues report that the obesity-associated noncoding sequences within *FTO* are actually functionally connected, at megabase distances, with *IRX3* [80]. Using so-called '3C' (chromatin conformation capture) technology, they show that sequences in intron 1 of *FTO* directly interacts with not only the *FTO* promoters, but also those of *IRX3* in both humans and mice, suggesting that the obesity-associated interval belongs to the regulatory landscape of *IRX3*. They then show that the *FTO* intron 1 risk SNPs are associated with expression of *IRX3*, but not *FTO*, in human brains [80]. The two caveats to this study being that (a) this expression work was done in post-mortem cerebella, as opposed to other brain regions more relevant to energy homeostasis and (b) that the differences in expression of *IRX3* between those carriers and non-carriers of the risk SNPs were very subtle indeed. The authors do however demonstrate a direct link between IRX3 expression and regulation of body mass and composition, with Irx3-deficient mice being 30 % smaller than their WT littermates, with reduced fat mass and increase in basal metabolic rate with browning of white adipose tissue. Additionally, hypothalamic expression of a dominant negative form of Irx3 reproduces the metabolic phenotypes of Irx3- deficient mice. Based on their findings, the authors conclude that the gene 'responsible' for the association of the SNPs to increased body-weight is *IRX3* and not *FTO* [80].

Where is the Smoking (Gene) Gun?

So is it all a fait accompli? Have we all been wasting our time and working on the 'wrong' gene? Not so fast there.

All of the findings I have detailed above regarding the biology of *FTO* are based on perturbation of its expression or on the use of recombinant protein, and is therefore independent of the risk SNPs in intron 1. It is unequivocal that *FTO* has a function in regulating body-size [65, 66, 68], likely a consequence of its role linking levels of amino acids to growth and translation [71, 72]. Additionally, there is evidence for co-regulatory mechanisms between *FTO* and *RPGRIP1L* with a possible overlapping regulatory region within intron 1 of *FTO* that contains at least two putative transcription factor binding sites (*CUX1*), as mentioned above, one of which overlaps with other obesity associated SNPs [45, 81].

As in Agatha Christie's 'Murder on the Orient Express', where all 12 passengers on the coach were complicit in the crime, it appears that a number of the genes in the region, certainly *FTO* and *IRX3*, play important roles in the regulation of body composition and size. It still remains a possibility that the association between *FTO* SNPs and body weight regulation is mediated through changing the expression of multiple genes in the 16q22 region.

In closing, we believe that the efforts to date in turning the statistical association of SNPs on 16q22 with increased BMI into a deeper understanding of a number of genes in the whole locus, could form a template into how we approach the many other emerging GWAS obesity SNPs, the majority of which are close to genes of unknown function.

References

1. Stunkard AJ, Foch TT, Hrubec Z (1986) A twin study of human obesity. JAMA 256:51–54
2. Stunkard AJ, Harris JR, Pedersen NL, McClearn GE (1990) The body-mass index of twins who have been reared apart. N Engl J Med 322:1483–1487
3. Yeo GS, Heisler LK (2012) Unraveling the brain regulation of appetite: lessons from genetics. Nat Neurosci 15:1343–1349
4. Frayling TM, Timpson NJ, Weedon MN, Zeggini E, Freathy RM, Lindgren CM et al (2007) A common variant in the *FTO* gene is associated with body mass index and predisposes to childhood and adult obesity. Science 316:889–894
5. Scuteri A, Sanna S, Chen WM, Uda M, Albai G, Strait J et al (2007) Genome-wide association scan shows genetic variants in the *FTO* gene are associated with obesity-related traits. PLoS Genet 3:e115
6. Dina C, Meyre D, Gallina S, Durand E, Korner A, Jacobson P et al (2007) Variation in *FTO* contributes to childhood obesity and severe adult obesity. Nat Genet 39:724–726
7. Berndt SI, Gustafsson S, Magi R, Ganna A, Wheeler E, Feitosa MF et al (2013) Genome-wide meta-analysis identifies 11 new loci for anthropometric traits and provides insights into genetic architecture. Nat Genet 45:501–512
8. Bradfield JP, Taal HR, Timpson NJ, Scherag A, Lecoeur C, Warrington NM et al (2012) A genome-wide association meta-analysis identifies new childhood obesity loci. Nat Genet 44:526–531
9. Lindgren CM, Heid IM, Randall JC, Lamina C, Steinthorsdottir V, Qi L et al (2009) Genome-wide association scan meta-analysis identifies three Loci influencing adiposity and fat distribution. PLoS Genet 5:e1000508
10. Meyre D, Delplanque J, Chevre JC, Lecoeur C, Lobbens S, Gallina S et al (2009) Genome-wide association study for early-onset and morbid adult obesity identifies three new risk loci in European populations. Nat Genet 41:157–159
11. Scherag A, Dina C, Hinney A, Vatin V, Scherag S, Vogel CI et al (2010) Two new Loci for body-weight regulation identified in a joint analysis of genome-wide association studies for early-onset extreme obesity in French and German study groups. PLoS Genet 6:e1000916

12. Speliotes EK, Willer CJ, Berndt SI, Monda KL, Thorleifsson G, Jackson AU et al (2010) Association analyses of 249,796 individuals reveal 18 new loci associated with body mass index. Nat Genet 42:937–948
13. Thorleifsson G, Walters GB, Gudbjartsson DF, Steinthorsdottir V, Sulem P, Helgadottir A et al (2009) Genome-wide association yields new sequence variants at seven loci that associate with measures of obesity. Nat Genet 41:18–24
14. Wheeler E, Huang N, Bochukova EG, Keogh JM, Lindsay S, Garg S et al (2013) Genome-wide SNP and CNV analysis identifies common and low-frequency variants associated with severe early-onset obesity. Nat Genet 45:513–517
15. Willer CJ, Speliotes EK, Loos RJ, Li S, Lindgren CM, Heid IM, Berndt SI et al (2009) Six new loci associated with body mass index highlight a neuronal influence on body weight regulation. Nat Genet 41:25–34
16. Cho YS, Go MJ, Kim YJ, Heo JY, Oh JH, Ban HJ et al (2009) A large-scale genome-wide association study of Asian populations uncovers genetic factors influencing eight quantitative traits. Nat Genet 41:527–534
17. Okada Y, Kubo M, Ohmiya H, Takahashi A, Kumasaka N, Hosono N, Maeda S et al (2012) Common variants at CDKAL1 and KLF9 are associated with body mass index in east Asian populations. Nat Genet 44:302–306
18. Wen W, Cho YS, Zheng W, Dorajoo R, Kato N, Qi L et al (2012) Meta-analysis identifies common variants associated with body mass index in east Asians. Nat Genet 44:307–311
19. Loos RJ, Yeo GS (2014) The bigger picture of *FTO*–the first GWAS-identified obesity gene. Nat Rev Endocrinol 10:51–61
20. Adeyemo A, Chen G, Zhou J, Shriner D, Doumatey A, Huang H, Rotimi C (2010) *FTO* genetic variation and association with obesity in West Africans and African Americans. Diabetes 59:1549–1554
21. Hassanein MT, Lyon HN, Nguyen TT, Akylbekova EL, Waters K, Lettre G et al (2010) Fine mapping of the association with obesity at the *FTO* locus in African-derived populations. Hum Mol Genet 19:2907–2916
22. Peters U, North KE, Sethupathy P, Buyske S, Haessler J, Jiao S, Fesinmeyer MD et al (2013) A systematic mapping approach of 16q12.2/*FTO* and BMI in more than 20,000 African Americans narrows in on the underlying functional variation: results from the Population Architecture using Genomics and Epidemiology (PAGE) study. PLoS Genet 9:e1003171
23. Monda KL, Chen GK, Taylor KC, Palmer C, Edwards TL, Lange LA et al (2013) A meta-analysis identifies new loci associated with body mass index in individuals of African ancestry. Nat Genet 45:690–696
24. Horikoshi M, Yaghootkar H, Mook-Kanamori DO, Sovio U, Taal HR, Hennig BJ et al (2013) New loci associated with birth weight identify genetic links between intrauterine growth and adult height and metabolism. Nat Genet 45:76–82
25. Jess T, Zimmermann E, Kring SI, Berentzen T, Holst C, Toubro S, Astrup A, Hansen T, Pedersen O, Sorensen TI (2008) Impact on weight dynamics and general growth of the common *FTO* rs9939609: a longitudinal Danish cohort study. Int J Obes (Lond) 32:1388–1394
26. Kilpelainen TO, den Hoed M, Ong KK, Grontved A, Brage S, Early Growth Genetics Consortium et al (2011) Obesity-susceptibility loci have a limited influence on birth weight: a meta-analysis of up to 28,219 individuals. Am J Cli Nutr 93:851–860
27. Graff M, Ngwa JS, Workalemahu T, Homuth G, Schipf S, Teumer A et al (2013) Genome-wide analysis of BMI in adolescents and young adults reveals additional insight into the effects of genetic loci over the life course. Hum Mol Genet 22:3597–3607
28. Hardy R, Wills AK, Wong A, Elks CE, Wareham NJ, Loos RJ, Kuh D, Ong KK (2010) Life course variations in the associations between *FTO* and MC4R gene variants and body size. Hum Mol Genet 19:545–552
29. Sovio U, Mook-Kanamori DO, Warrington NM, Lawrence R, Briollais L, Palmer CN et al (2011) Association between common variation at the *FTO* locus and changes in body mass index from infancy to late childhood: the complex nature of genetic association through growth and development. PLoS Genet 7:e1001307

30. Cecil JE, Tavendale R, Watt P, Hetherington MM, Palmer CN (2008) An obesity-associated *FTO* gene variant and increased energy intake in children. N Engl J Med 359:2558–2566
31. Speakman JR, Rance KA, Johnstone AM (2008) Polymorphisms of the *FTO* gene are associated with variation in energy intake, but not energy expenditure. Obesity (Silver Spring) 16:1961–1965
32. Timpson NJ, Emmett PM, Frayling TM, Rogers I, Hattersley AT, McCarthy MI, Davey Smith G (2008) The fat mass- and obesity-associated locus and dietary intake in children. Am J Clin Nutr 88:971–978
33. Park SL, Cheng I, Pendergrass SA, Kucharska-Newton AM, Lim U, Ambite JL et al (2013) Association of the *FTO* obesity risk variant rs8050136 with percentage of energy intake from fat in multiple racial/ethnic populations: the PAGE study. Am J Epidemiol 178:780–790
34. Sonestedt E, Roos C, Gullberg B, Ericson U, Wirfalt E, Orho-Melander M (2009) Fat and carbohydrate intake modify the association between genetic variation in the *FTO* genotype and obesity. Am J Clin Nutr 90:1418–1425
35. Wardle J, Carnell S, Haworth CM, Farooqi IS, O'Rahilly S, Plomin R (2008) Obesity associated genetic variation in *FTO* is associated with diminished satiety. J Clin Endocrinol Metab 93:3640–3643
36. Wardle J, Llewellyn C, Sanderson S, Plomin R (2009) The *FTO* gene and measured food intake in children. Int J Obes (Lond) 33:42–45
37. Tanofsky-Kraff M, Han JC, Anandalingam K, Shomaker LB, Columbo KM, Wolkoff LE et al (2009) The *FTO* gene rs9939609 obesity-risk allele and loss of control over eating. Am J Clin Nutr 90:1483–1488
38. Tanaka T, Ngwa JS, van Rooij FJ, Zillikens MC, Wojczynski MK, Frazier-Wood AC et al (2013) Genome-wide meta-analysis of observational studies shows common genetic variants associated with macronutrient intake. Am J Clin Nutr 97:1395–1402
39. Ahmad T, Chasman DI, Mora S, Pare G, Cook NR, Buring JE, Ridker PM, Lee IM (2010) The fat-mass and obesity-associated (*FTO*) gene, physical activity, and risk of incident cardiovascular events in white women. Am Heart J 160:1163–1169
40. Franks PW, Jablonski KA, Delahanty LM, McAteer JB, Kahn SE, Knowler WC et al (2008) Assessing gene-treatment interactions at the *FTO* and INSIG2 loci on obesity-related traits in the Diabetes Prevention Program. Diabetologia 51:2214–2223
41. Vimaleswaran KS, Li S, Zhao JH, Luan J, Bingham SA, Khaw KT, Ekelund U, Wareham NJ, Loos RJ (2009) Physical activity attenuates the body mass index-increasing influence of genetic variation in the *FTO* gene. Am J Clin Nutr 90:425–428
42. Kilpelainen TO, Qi L, Brage S, Sharp SJ, Sonestedt E, Demerath E et al (2011) Physical activity attenuates the influence of *FTO* variants on obesity risk: a meta-analysis of 218,166 adults and 19,268 children. PLoS Med 8:e1001116
43. Yang J, Loos RJ, Powell JE, Medland SE, Speliotes EK, Chasman DI et al (2012) *FTO* genotype is associated with phenotypic variability of body mass index. Nature 490:267–272
44. Cantor RM, Lange K, Sinsheimer JS (2010) Prioritizing GWAS results: a review of statistical methods and recommendations for their application. Am J Hum Genet 86:6–22
45. Stratigopoulos G, Padilla SL, LeDuc CA, Watson E, Hattersley AT, McCarthy MI, Zeltser LM, Chung WK, Leibel RL (2008) Regulation of *FTO*/Ftm gene expression in mice and humans. Am J Physiol Regul Integr Comp Physiol 294:R1185–1196
46. Gerken T, Girard CA, Tung YC, Webby CJ, Saudek V, Hewitson KS et al (2007) The obesity-associated *FTO* gene encodes a 2-oxoglutarate-dependent nucleic acid demethylase. Science 318:1469–1472
47. Tung YC, Ayuso E, Shan X, Bosch F, O'Rahilly S, Coll AP, Yeo GS (2010) Hypothalamic-specific manipulation of *FTO*, the ortholog of the human obesity gene *FTO*, affects food intake in rats. PloS ONE 5:e8771
48. Delous M, Baala L, Salomon R, Laclef C, Vierkotten J, Tory K et al (2007) The ciliary gene RPGRIP1L is mutated in cerebello-oculo-renal syndrome (Joubert syndrome type B) and Meckel syndrome. Nat Genet 39:875–881

49. Jia G, Yang CG, Yang S, Jian X, Yi C, Zhou Z, He C (2008) Oxidative demethylation of 3-methylthymine and 3-methyluracil in single-stranded DNA and RNA by mouse and human *FTO*. FEBS Lett 582:3313–3319
50. Jia G, Fu Y, Zhao X, Dai Q, Zheng G, Yang Y et al (2011) N6-methyladenosine in nuclear RNA is a major substrate of the obesity-associated *FTO*. Nat Chem Biol 7:885–887
51. Han Z, Niu T, Chang J, Lei X, Zhao M, Wang Q, Cheng W, Wang J, Feng Y, Chai J (2010) Crystal structure of the *FTO* protein reveals basis for its substrate specificity. Nature 464:1205–1209
52. Meyre D, Proulx K, Kawagoe-Takaki H, Vatin V, Gutierrez-Aguilar R, Lyon D et al (2010) Prevalence of loss-of-function *FTO* mutations in lean and obese individuals. Diabetes 59:311–318
53. Desrosiers R, Friderici K, Rottman F (1974) Identification of methylated nucleosides in messenger RNA from Novikoff hepatoma cells. Proc Natl Acad Sci U S A 71:3971–3975
54. Kowalak JA, Pomerantz SC, Crain PF, McCloskey JA (1993) A novel method for the determination of post-transcriptional modification in RNA by mass spectrometry. Nucleic Acids Res 21:4577–4585
55. Challis BG, Pritchard LE, Creemers JW, Delplanque J, Keogh JM, Luan J et al (2002) A missense mutation disrupting a dibasic prohormone processing site in pro-opiomelanocortin (POMC) increases susceptibility to early-onset obesity through a novel molecular mechanism. Hum Mol Genet 11:1997–2004
56. Krude H, Biebermann H, Luck W, Horn R, Brabant G, Gruters A (1998) Severe early-onset obesity, adrenal insufficiency and red hair pigmentation caused by POMC mutations in humans. Nat Genet 19:155–157
57. Benzinou M, Creemers JW, Choquet H, Lobbens S, Dina C, Durand E et al (2008) Common nonsynonymous variants in PCSK1 confer risk of obesity. Nat Genet 40:943–945
58. Jackson RS, Creemers JW, Ohagi S, Raffin-Sanson ML, Sanders L, Montague CT, Hutton JC, O'Rahilly S (1997) Obesity and impaired prohormone processing associated with mutations in the human prohormone convertase 1 gene. Nat Genet 16:303–306
59. Loos RJ, Lindgren CM, Li S, Wheeler E, Zhao JH, Prokopenko I et al (2008) Common variants near MC4R are associated with fat mass, weight and risk of obesity. Nat Genet 40:768–775
60. Yeo GS, Farooqi IS, Aminian S, Halsall DJ, Stanhope RG, O'Rahilly S (1998) A frameshift mutation in MC4R associated with dominantly inherited human obesity. Nat Genet 20:111–112
61. Gray J, Yeo GS, Cox JJ, Morton J, Adlam AL, Keogh JM et al (2006) Hyperphagia, severe obesity, impaired cognitive function, and hyperactivity associated with functional loss of one copy of the brain-derived neurotrophic factor (BDNF) gene. Diabetes 55:3366–3371
62. Peters T, Ausmeier K, Ruther U (1999) Cloning of Fatso (*FTO*), a novel gene deleted by the Fused toes (Ft) mouse mutation. Mamm Genome 10:983–986
63. Anselme I, Laclef C, Lanaud M, Ruther U, Schneider-Maunoury S (2007) Defects in brain patterning and head morphogenesis in the mouse mutant Fused toes. Dev biol 304:208–220
64. van der Hoeven F, Schimmang T, Volkmann A, Mattei MG, Kyewski B, Ruther U (1994) Programmed cell death is affected in the novel mouse mutant Fused toes (Ft). Development 120:2601–2607
65. Fischer J, Koch L, Emmerling C, Vierkotten J, Peters T, Bruning JC, Ruther U (2009) Inactivation of the *FTO* gene protects from obesity. Nature 458:894–898
66. McMurray F, Church CD, Larder R, Nicholson G, Wells S, Teboul L et al (2013) Adult onset global loss of the *FTO* gene alters body composition and metabolism in the mouse. PLoS Genet 9:e1003166
67. Boissel S, Reish O, Proulx K, Kawagoe-Takaki H, Sedgwick B, Yeo GS et al (2009) Loss-of-function mutation in the dioxygenase-encoding *FTO* gene causes severe growth retardation and multiple malformations. Am J Hum Genet 85:106–111
68. Church C, Moir L, McMurray F, Girard C, Banks GT, Teboul L et al (2010) Overexpression of *FTO* leads to increased food intake and results in obesity. Nat Genet 42:1086–1092

69. Gao X, Shin YH, Li M, Wang F, Tong Q, Zhang P (2010) The fat mass and obesity associated gene *FTO* functions in the brain to regulate postnatal growth in mice. PloS ONE 5:e14005
70. Ma M, Harding HP, O'Rahilly S, Ron D, Yeo GS (2012) Kinetic analysis of FTO (fat mass and obesity-associated) reveals that it is unlikely to function as a sensor for 2-oxoglutarate. Biochem J 444(2):183–187
71. Cheung MK, Gulati P, O'Rahilly S, Yeo GS (2013) FTO expression is regulated by availability of essential amino acids. Int J Obes (Lond) 37(5):744–747
72. Gulati P, Cheung MK, Antrobus R, Church CD, Harding HP, Tung YC, Rimmington D, Ma M, Ron D, Lehner PJ, Ashcroft FM, Cox RD, Coll AP, O'Rahilly S, Yeo GS (2013) Role for the obesity-related FTO gene in the cellular sensing of amino acids. Proc Natl Acad Sci U S A 110(7):2557–2562
73. Quevillon S, Robinson JC, Berthonneau E, Siatecka M, Mirande M (1999) Macromolecular assemblage of aminoacyl-tRNA synthetases: identification of protein-protein interactions and characterization of a core protein. J Mol Biol 285(1):183–195
74. Dominissini D, Moshitch-Moshkovitz S, Schwartz S, Salmon-Divon M, Ungar L, Osenberg S, Cesarkas K, Jacob-Hirsch J, Amariglio N, Kupiec M, Sorek R, Rechavi G (2012) Topology of the human and mouse m6A RNA methylomes revealed by m6A-seq. Nature 485(7397):201–206
75. Meyer KD, Saletore Y, Zumbo P, Elemento O, Mason CE, Jaffrey SR (2012) Comprehensive analysis of mRNA methylation reveals enrichment in 3' UTRs and near stop codons. Cell 149(7):1635–1646
76. Bokar JA, Shambaugh ME, Polayes D, Matera AG, Rottman FM (1997) Purification and cDNA cloning of the AdoMet-binding subunit of the human mRNA (N6-adenosine)-methyltransferase. RNA 3(11):1233–1247
77. Zheng G, Dahl JA, Niu Y, Fedorcsak P, Huang CM, Li CJ, Vågbø CB, Shi Y, Wang WL, Song SH, Lu Z, Bosmans RP, Dai Q, Hao YJ, Yang X, Zhao WM, Tong WM, Wang XJ, Bogdan F, Furu K, Fu Y, Jia G, Zhao X, Liu J, Krokan HE, Klungland A, Yang YG, He C (2013) ALKBH5 is a mammalian RNA demethylase that impacts RNA metabolism and mouse fertility. Mol Cell 49(1):18–29
78. Hess ME, Hess S, Meyer KD, Verhagen LA, Koch L, Brönneke HS, Dietrich MO, Jordan SD, Saletore Y, Elemento O, Belgardt BF, Franz T, Horvath TL, Rüther U, Jaffrey SR, Kloppenburg P, Brüning JC (2013) The fat mass and obesity associated gene (Fto) regulates activity of the dopaminergic midbrain circuitry. Nat Neurosci 16(8):1042–1048
79. Karra E, O'Daly OG, Choudhury AI, Yousseif A, Millership S, Neary MT, Scott WR, Chandarana K, Manning S, Hess ME, Iwakura H, Akamizu T, Millet Q, Gelegen C, Drew ME, Rahman S, Emmanuel JJ, Williams SC, Rüther UU, Brüning JC, Withers DJ, Zelaya FO, Batterham RL (2013) A link between FTO, ghrelin, and impaired brain food-cue responsivity. J Clin Invest 123(8):3539–3551
80. Smemo S, Tena JJ, Kim KH, Gamazon ER, Sakabe NJ, Gómez-Marín C et al (2014) Obesity-associated variants within FTO form long-range functional connections with IRX3. Nature 507(7492):371–375
81. Stratigopoulos G, LeDuc CA, Cremona ML, Chung WK, Leibel RL (2011) Cut-like homeobox 1 (CUX1) regulates expression of the fat mass and obesity-associated and retinitis pigmentosa GTPase regulator-interacting protein-1-like (RPGRIP1L) genes and coordinates leptin receptor signaling. J Biol Chem 286(3):2155–2170

Chapter 6
Neural Vulnerability Factors that Increase Risk for Weight Gain: Prevention and Treatment Implications

Eric Stice, Sonja Yokum and Kyle Burger

Introduction

Nearly 70% of US adults are overweight or obese [1], which increases risk for coronary heart disease, atherosclerotic cerebrovascular disease, colorectal cancer, and all-cause mortality, is credited with 300,000 US deaths and $ 150 billion in health-related expenses yearly [2, 3]. Unfortunately, treatments almost never result in lasting weight loss and virtually all obesity prevention programs have not reduced future obesity onset [4, 5]. An improved understanding of the risk processes that give rise to weight gain should guide the design of more effective preventive programs and treatments. At present, most risk factors that predict future weight gain show only weak effects [6–9]. For example, the predictive effects for parental obesity, a well replicated risk factor for future weight gain, have only ranged from an $r=0.18$ to 0.21 in large epidemiologic studies [e.g., 7, 9].

Theorists have focused on the role of reward circuitry in obesity because eating palatable food increases activation in regions implicated in reward, including the striatum, midbrain, amygdala, and orbitofrontal cortex (OFC) [10–12] and causes dopamine (DA) release in the dorsal striatum, with the amount released correlating with meal pleasantness ratings [13]. Indeed, even delivery of high-fat food directly to the gut, bypassing the oral cavity, has been shown to induce robust striatal dopamine release in rodents [14]. It has been posited that aberrant reward-related responses to food intake and/or cues override homeostatic processes of hunger and fullness, resulting in excess adipose tissue and weight gain [e.g., 15]. Further, data indicate that appetitive hormones thought to influence homeostatic determinants of food intake (e.g., leptin, peptide YY, glucagon-like peptide 1) act by altering reward value of food [16]. In support of hedonically driven food intake, direct phar-

E. Stice (✉) · S. Yokum · K. Burger
Oregon Research Institute, Eugene, OR, USA
e-mail: estice@ori.org

C. Nóbrega, R. Rodriguez-López (eds.), *Molecular Mechanisms Underpinning the Development of Obesity,* DOI 10.1007/978-3-319-12766-8_6

macological activation of the striatum prompts hyperphagia in animals, increasing preferential intake of high-fat/sugar foods, even in sated animals [17]. Agonist and antagonist experiments suggest that DA signaling plays a larger role in reward learning and that opioid peptide signaling plays a larger role in hedonic pleasure from food intake [18], though reward regions contain both DA and opioid receptors and opioid agonists cause DA signaling [19], implying the two neurotransmitter systems are tightly intertwined.

Reward Surfeit and Incentive Sensitization Theories of Obesity

The reward surfeit model holds that individuals who showed greater responsivity of reward regions to food intake, which is presumably inborn, are at elevated risk for overeating and consequent weight gain [20]. The incentive sensitization model posits that repeated intake of palatable foods results in an elevated responsivity of reward valuation regions to cues that are repeatedly associated with palatable food intake via conditioning, which prompts elevated food intake when these cues are encountered [18].

Consistent with these theories, obese versus lean humans show greater responsivity of brain regions associated with reward and motivation (striatum, amygdala, OFC) to pictures of high-fat/sugar foods versus low-fat/sugar foods and control images [21–28] and to pictorial cues that signal impending palatable food receipt [20, 29]. These data are supported by studies examining acute and longer-term food intake. Specifically, midbrain and medial OFC activity in response to milkshake receipt positively predicted subsequent *ad libitum* milkshake consumption and BOLD response in the ventral striatum during exposure to food images positively predicts later snack consumption [30, 31]. Using objectively measured energy intake over a two-week period in lean adolescents, a positive relation was observed between energy intake beyond basal metabolic needs and BOLD response during cues predicting food receipt in regions thought to encode visual processing and attention (visual and anterior cingulate cortices), salience (precuneus), as well as the primary gustatory cortex (frontal operculum) and (reward/motivation) striatum [32]. Animal experiments indicate that firing of striatal and ventral pallidum DA neurons initially occurs in response to receipt of a novel palatable food, but that after repeated pairings of palatable food intake and cues that signal impending receipt of that food, DA neurons begin to fire in response to reward-predictive cues and no longer fire in response to food receipt [33–35]. Theorists posit this shift during cue-reward learning acts to either update knowledge regarding the predictive cues or attribute reward value to the cues themselves thereby guiding behavior [36–39].

Critically, fMRI studies indicate that hyper-responsivity of reward regions (striatum, amygdala, OFC) to palatable food images [40, 41], palatable food odors [42] or cues that signal impending presentation of palatable food images [43] predicts future weight gain. Additionally, teens that exhibit greater striatal response to high-fat/sugar food commercials show elevated weight gain over 1-year

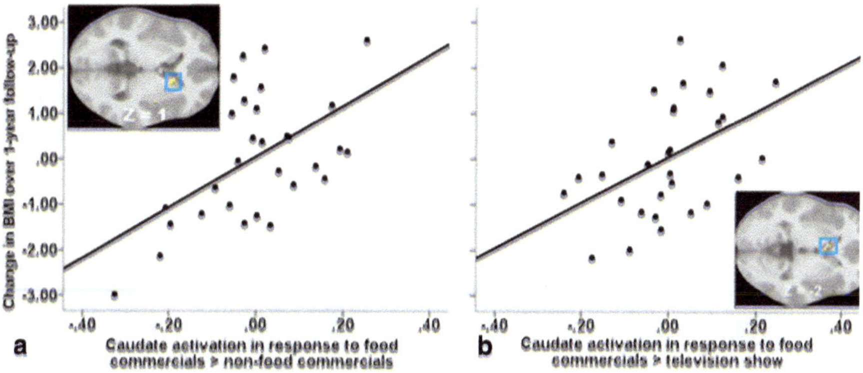

Fig. 6.1 Partial regression plots showing the relations of **a** activation in the caudate (MNI coordinates: 12, 14, 1) in response to food commercials > non-food commercials and **b** activation in the caudate (MNI coordinates: −9, 14, −2) in response to food commercials > television show to change in BMI over 1-year follow-up

follow-up (r = 0.47–0.51; Fig. 6.1; [44]). Elevated amygdala, midbrain, thalamus, hypothalamus, ventral pallidum, and nucleus accumbens responsivity to tastes of milkshake have also predicted weight gain over 1-year follow-up [42, 45].

Interestingly, there is evidence that the effects of hyper-responsivity of reward regions to food and food cues shows significantly stronger relations to future weight gain for individuals with a genetic propensity for greater DA signaling capacity in reward regions. Adolescents with elevated caudate and putamen responsivity to milkshake tastes who have a genetic propensity for greater DA signaling due to possessing an A2/A2 *TaqIA* allele showed significantly greater weight gain over 1-year follow-up [46]. Likewise, teens who show elevated striatal/OFC response to palatable food images and who had a genetic propensity for greater DA signaling due to possessing an A2/A2 *TaqIA* allele, also showed elevated future weight gain [27]. Similar effects have emerged for another genotype (no seven-repeat or longer alleles of the *DRD4* gene [*DRD4*-S]) that has been associated with elevated DA signaling [27]. Data from a large ($N = 155$) ongoing study revealed that lean teens who showed greater OFC response to a cue that signals impending milkshake receipt were more likely to gain weight over 2-year follow-up ($r = 0.29$) and that this relation was significantly stronger for youth with a genetic propensity for greater DA signaling capacity in reward circuitry as indexed by a multilocus composite that reflects the number of alleles associated with elevated DA signaling (Fig. 6.2). We examined this multilocus score because it relates more strongly to reward region responsivity than the individual alleles used to calculate this composite genetic risk score [47, 48]. Theoretically, this is because a greater number of these genotypes, regardless of the particular combination, are associated with greater DA signaling capacity. The multilocus composite was scored as follows: *TaqIA* A1/A1, *DRD2-141*C Ins/Ins, *DRD4*-L, *DAT1* 10R/10R, and *COMT* Met/Met genotypes were assigned a score of 0 ('low'); *TaqIA* A2/A2, *DRD2-141*C Ins/Del and Del/Del, *DRD4*-S, *DAT1* 9R, and *COMT* Val/Val genotypes were assigned a score of 1 ('high'), and

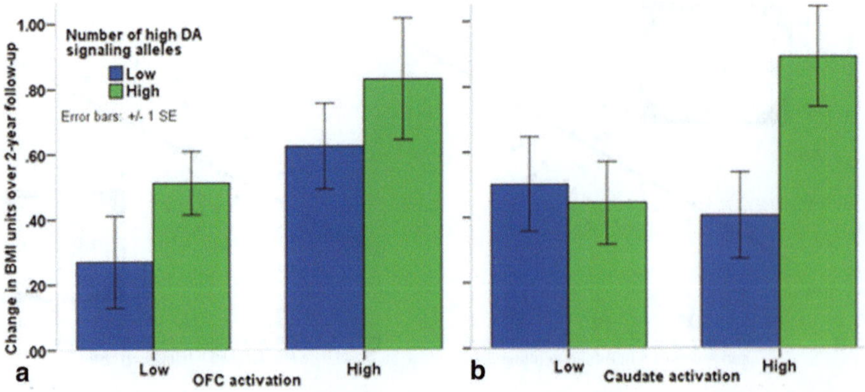

Fig. 6.2 Change in BMI predicted by the interaction between the number of high DA signaling alleles and **a** OFC activation and **b** caudate activation in response to anticipatory milkshake receipt

TaqIA A1/A2 and *COMT* Met/Val genotypes received a score of 0.5. Scores were summed to create the composite. Humans with the A2/A2 allele versus an A1 allele of the *Taq1A* polymorphism and the Del allele versus Ins/Ins genotype of the *DRD2-141*C Ins/Del polymorphism show more D2 receptors [49]. Humans with the shorter than seven alleles (*DRD4*-S) versus seven-repeat or longer allele (*DRD4*-L) of the *DRD4* genotype show greater *in vitro* DA functioning and stronger response to DA agonists [50, 51]. Humans with the nine-repeat allele (*DAT1-S)* versus homozygous for the ten-repeat allele (*DAT1*-L) of the *DAT1* show lower *DAT1* expression [52], theoretically increasing synaptic DA clearance, producing lower basal DA levels and increased phasic DA release [53]. Val homozygotes versus Met homozygotes of the *Catechol-O-methyltransferase* (*COMT* val[158]met) gene putatively have lower basal striatal DA levels and greater phasic DA release [54]. Individuals with higher multilocus scores showed greater future weight gain in three separate studies (Fig. 6.3; [55]), confirming that this effect is replicable.

Thus, studies from multiple independent labs have found that individuals who show elevated reward region responsivity to palatable food intake are more likely to enter a prolonged period of positive energy balance and gain weight, providing key behavioral data in support of the reward surfeit theory of obesity. Studies from multiple independent labs have also found that individuals who show elevated reward region responsivity to cues that have been associated with palatable food intake show elevated future weight gain, providing behavioral support for the incentive sensitization theory of obesity. There was also evidence that the predictive relations of elevated reward region responsivity to palatable food intake and food cues to future weight gain are stronger for individuals with a genetic propensity for greater DA signaling capacity in reward regions, which might be construed as further support for the reward surfeit model. These results imply that reducing habitual intake of high-fat and high-sugar foods should theoretically reduce the conditioning process that leads to elevated reward region responsivity to food cues, which may be an

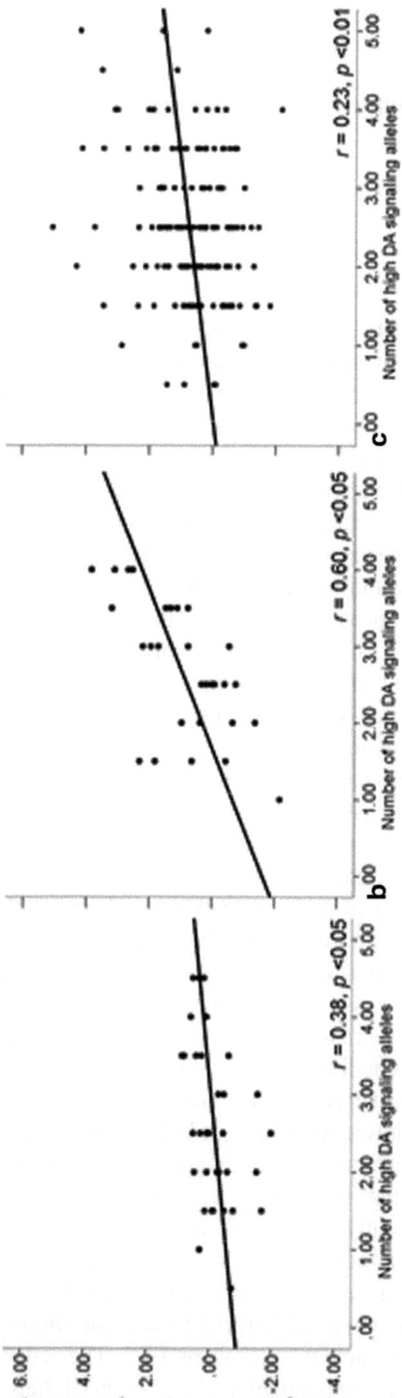

Fig. 6.3 Partial regression plots showing the effects of the number of high DA signaling alleles (multilocus composite score) on change in BMI over 1-year follow-up in **a** 38 young females (*M* age=20.9±1.2; *M* BMI=28.1; BMI range=24.0–39.0), **b** 30 adolescents (56.7 % female; *M* age=15.2±1.0; *M* BMI=26.9; BMI range=19.5–37.9), and **c** 161 adolescents (50.6 % female; *M* age=15.2±1.0; *M* BMI=20.8; BMI range=17.0–26.1), while controlling for baseline BMI

effective method of reducing risk for weight gain. However, the fact that behavioral weight loss programs typically result in reduced intake of such foods, typically leading to a 10 % weight loss on average, but do not produced sustained weight loss implies that it is very difficult to reduce reward region hyper-responsivity to food cues once it has emerged.

Reward Deficit Theory of Obesity

The reward deficit model of obesity posits that individuals with lower sensitivity of DA-based reward regions overeat to compensate for this deficiency [56]. Apparently consistent with this theory, obese versus lean adults show lower striatal DA D2 receptor availability [57–59], though two other studies found no significant group differences [60, 61], which might have been due to the smaller samples sizes in the latter studies. Obese versus lean adults show lower capacity of nigrostriatal neurons to synthesize DA [62], and less striatal responsivity to tastes of high-fat/sugar beverages [20, 46, 63–65]. Obese versus lean rats likewise have lower basal DA levels and D2 receptor availability and less ex vivo DA release in response to electrical stimulation in nucleus accumbens and dorsal striatum tissue [66–69]. Interestingly, a recent study in humans [58] found a positive correlation between BMI and DA release in the dorsal striatum and substantia nigra in response to amphetamine, suggesting that D2 receptor availability may not be closely coupled with degree of DA response from rewarding experiences.

Although the above cross-sectional findings appear to provide some support for the reward deficit theory of obesity, prospective and experimental findings indicate that overeating contributes to reward region hypo-responsivity. Lean youth at risk for future obesity by virtue of parental obesity show hyper-responsivity of reward regions to palatable food receipt and monetary reward, rather than hypo-responsivity [70]. Young women who gained weight over a 6-month period showed a reduction in striatal responsivity to palatable food receipt relative to women who remained weight stable [71]. This finding converges with experimental overfeeding studies involving animals; rats randomized to overeating conditions that results in weight gain versus control conditions show down-regulation of post-synaptic D2 receptors, and reduced D2 sensitivity, extracellular DA levels in the nucleus accumbens and DA turnover, and lower sensitivity of DA reward circuitry to food intake, electrical stimulation, amphetamine administration, and potassium administration [69, 72–76]. Of note, rats that had consumed a high-fat/sugar diet continued to eat that food when subsequently paired with foot shocks, whereas chow-diet rats would not eat high-fat/sugar food when paired with foot shocks [76], suggesting that energy dense diets may induce compulsive eating. Further, pigs randomized to a weight gain intervention versus a stable weight condition showed reduced resting activity in the midbrain and nucleus accumbens [77]. The reduced DA signaling capacity appears to occur because habitual intake of high-fat diets causes decreased synthesis of oleoylethanolamine, a gastrointestinal lipid messenger [14]. Further, people who report elevated intake of particular foods show reduced striatal response during intake of that food, independent of BMI [78–80].

Given that animals that habitually use drugs of abuse that produce similar down-regulation of reward regions will work to keep DA levels in the nucleus accumbens above a certain level [81–83], Geiger and associated [74] speculate that rats which have experienced diet-induced down-regulation of DA circuitry may similarly overeat to increase DA signaling. However, a more recent study found that mice in which reduced striatal DA signaling from food intake was experimentally induced through intragastric feeding of high-fat food worked less for intragastric administration of high-fat food and consumed less food *ad lib* than control mice [14]. These experimental results seem incompatible with the notion that an induced down-regulation of DA reward circuitry leads to compensatory overeating. Results from the Tellez et al. [14] study also provided further evidence that intake of high-fat food can result in reduced DA response to food intake, independent of weight gain per se.

Interestingly, the relations between lower striatal response to milkshake receipt and weight gain over 1-year follow-up [20] and between lower putamen and OFC response to palatable food images and weight gain over 1-year follow-up [27] were significantly stronger for youth with the A1 allele, which is associated with less DA signaling, implying that any reduction in DA signaling caused by overeating may have a more pronounced reward deficit effect for those at genetic risk for lower DA signaling. Similar effects have emerged for individuals with the seven-repeat allele of the *DRD4* gene, which is also associated with reduced DA signaling capacity [27].

Thus, studies have provided little prospective or experimental support for the reward deficit theory of obesity. Specifically, no prospective study has reported a main effect between reduced reward region responsivity to food intake or cues and future weight gain. Indeed, as noted, prospective studies have found that greater responsivity of reward circuitry, including the amygdala, midbrain, ventral pallidum, nucleus accumbens, and striatal, rather than reduced responsivity, to palatable milkshake intake predicts future weight gain [e.g., 42, 45]. And recent data found that inducing down-regulation of DA response to food intake resulted in less caloric intake and motivation for food than observed in control mice [14]. Thus, findings collectively suggest that the reduced DA signaling capacity of reward circuitry can be acquired from overeating, but provide little support for the notion that this contributes to overeating and subsequent weight gain.

Translating Findings from Brain Imaging Research into an Obesity Prevention Program

Thus, emerging research suggests that obese versus lean individuals show elevated reward region responsivity to images of high-fat/high-sugar foods and that this increases risk for future weight gain. Fortunately, there is evidence that prefrontal regions can reduce reward region responsivity to appetitive cues [84]. Cognitive reappraisals, such as thinking of the long-term health consequences of eating unhealthy food when viewing images of such foods, increases inhibitory region (dlPFC, vlPFC, vmPFC, lateral OFC, superior and inferior frontal gyrus) activation and decreases reward region (ventral striatum, amygdala, ACC, VTA, posterior insula) and attention region (precuneus, posterior cingulate cortex) activation relative to contrast conditions, such as imagined intake [85–88]. Stoeckel and associates [89] used real-time fMRI biofeedback to augment the effects of cognitive reappraisals in reducing reward region responsivity and increasing inhibitory region responsivity to images of palatable foods; the training resulted in significantly greater reduction in medial OFC, right ventral striatum, and right amygdala, as well as greater activation in an inhibitory control region (inferior frontal cortex [IFG]) in response to palatable food images. Thus, findings suggest that cognitive reappraisals may reduce hyper-responsivity of reward regions to food cues and increase inhibitory control region activation, which is crucial because our environment is replete with food images and cues (e.g., ads on TV) that contribute to overeating. For instance, US teens are exposed to over 5000 unhealthy food commercials yearly [90]. Indeed, exposure to unhealthy food commercials results in greater caloric intake of the advertised foods and other unhealthy foods [90–92]. Accordingly, we developed an obesity prevention program that trained participants to use cognitive reappraisals when confronted with unhealthy tempting foods. We hypothesize that if participants learn to automatically apply these cognitive reappraisals, they will show reduced reward and attention region responsivity and increased inhibitory region responsivity to food images and cues signaling impending delivery of a high-fat/high-sugar food, which should result in reduced caloric intake and weight gain.

Emerging data also suggests that obese versus lean individuals show reduced recruitment of reward regions during intake of high-fat/high-sugar foods, which is either inborn of acquired, with some evidence that this may increase risk for future weight gain. If overeating energy-dense food reduces reward region response to such food, which may prompt compensatory overeating to achieve the same satisfaction experienced previously, reducing fat and sugar intake may help people avoid this induced-reward deficit that may contribute to obesity. Such a "palate-retraining" intervention may also reduce preferences for high-fat/sugar foods, which may contribute to weight gain. Reducing intake of dietary fat decreases preferences and frequency of future consumption of previously preferred high-fat foods and increases acceptance of low-fat foods [93, 94], implying a relation between habitual fat intake and preferences for fat foods. Chronic intake of a high-fat diet theoretically leads to reduced oral sensitivity, prompting compensatory escalations in fat intake to experience same degree of reward [95]. We therefore included a palate-retraining component to our neuroimaging-informed obesity prevention program wherein participants reduce fat and sugar intake to decrease taste preferences for high-fat/sugar foods and avoid the reduced reward region responsivity to high-fat/sugar food intake observed in obese humans. We hypothesize that if intervention participants reduce overall consumption of fat and sugar, they will show an increase in striatal response to receipt of a high-fat/high-sugar milkshake, which may reduce risk for compensatory overeating.

We therefore evaluated an obesity prevention program that trained young adults to (a) use cognitive reappraisals when confronted with tempting palatable foods and (b) gradually reduce intake of fat and sugar in their diets to decrease risk for a blunted striatal response to palatable food intake [96]. Young adults at risk for future weight gain by virtue of weight concerns (N=148) were randomized to this new *Minding Health* prevention program, an alternative prevention program promoting participant-driven gradual reductions in caloric intake and increases in physical activity (the *Healthy Weight* intervention), or an obesity education video control condition, completing assessments at pre, post, and 6-month follow-up. A subset of *Minding Health* and control participants completed an fMRI scan at pre and post assessing neural response to images of high-fat/sugar foods and to receipt and anticipated receipt of a high-fat/sugar food. *Minding Health* participants showed significantly greater reductions in body fat than controls and percentage of caloric intake from fat and sugar than *Healthy Weight* participants, though these effects attenuated somewhat by 6-month follow-up. However, *Healthy Weight* participants showed greater reductions in BMI and eating disorder symptoms than *Minding Health* participants. *Minding Health* participants showed greater activation of an inhibitory control region (IFG) and reduced activation of an attention/expectation region (mid cingulate gyrus) in response to palatable food images relative to pretest and controls. Although the *Minding Health* intervention produced some of the hypothesized effects, it only affected some outcomes and the effects often showed limited persistence.

Future Research Directions

This review highlights several important directions for future research. First, although a small number of prospective brain imaging studies have investigated neural vulnerability factors that predict future weight gain, the vast majority of the literature reviewed herein is cross-sectional. It will therefore be important for additional large sample prospective brain imaging studies to identify neural vulnerability factors that predict future weight gain. Second, it will be important to investigate environmental, social, and biological factors that amplify and mitigate the effects of these vulnerability factors on future weight gain. Third, it would be useful for additional prospective repeated-measures studies to attempt to capture the plasticity of reward region responsivity to food images/cues and food receipt, that appears to emerge secondary to overeating, which may play a role in maintaining overeating. If randomized experiments could be used to address these three directions for future research, much stronger inferences regarding these etiologic processes would be possible. Finally, we hope that future research will continue to try to translate findings from brain imaging studies into prevention and treatment interventions for obesity. For instance, we suspect it would be useful to test whether real-time fMRI biofeedback could be used to enhance the efficacy of the cognitive reappraisal-based obesity prevention program described above. It might likewise be possible to use non-invasive brain stimulation procedures to augment the efficacy of these types of obesity prevention programs.

References

1. Flegal KM, Carroll MD, Kit BK, Ogden CL (2012) Prevalence of obesity and trends in the distribution of body mass index among US adults, 1999–2010. JAMA 307(5):491–497
2. Finkelstein EA, Trogdon JG, Cohen JW, Dietz W (2009) Annual medical spending attributable to obesity: payer-and service-specific estimates. Health Aff (Millwood) 28(5):w822–831
3. Flegal KM, Graubard BI, Williamson DF, Gail MH (2005) Excess deaths associated with underweight, overweight, and obesity. JAMA 293(15):1861–1867
4. Stice E, Shaw H, Marti CN (2006) A meta-analytic review of obesity prevention programs for children and adolescents: the skinny on interventions that work. Psychol Bull 132(5):667–691
5. Turk MW, Yang K, Hravnak M, Sereika SM, Ewing LJ, Burke LE (2009) Randomized clinical trials of weight loss maintenance: a review. J Cardiovasc Nurs 24(1):58–80
6. Haines J, Neumark-Sztainer D, Wall M, Story M (2007) Personal, behavioral, and environmental risk and protective factors for adolescent overweight. Obesity (Silver Spring) 15(11):2748–2760
7. Salbe AD, DelParigi A, Pratley RE, Drewnowski A, Tataranni PA (2004) Taste preferences and body weight changes in an obesity-prone population. Am J Clin Nutr 79(3):372–378
8. Stice E, Presnell K, Shaw H, Rohde P (2005) Psychological and behavioral risk factors for obesity onset in adolescent girls: a prospective study. J Consult Clin Psychol 73(2):195–202
9. Whitaker RC, Wright JA, Pepe MS, Seidel KD, Dietz WH (1997) Predicting obesity in young adulthood from childhood and parental obesity. N Engl J Med 337(13):869–873
10. Kringelbach ML, O'Doherty J, Rolls ET, Andrews C (2003) Activation of the human orbitofrontal cortex to a liquid food stimulus is correlated with its subjective pleasantness. Cereb Cortex 13(10):1064–1071
11. Small DM, Zatorre RJ, Dagher A, Evans AC, Jones-Gotman M (2001) Changes in brain activity related to eating chocolate: from pleasure to aversion. Brain 124(Pt 9):1720–1733
12. Stice E, Burger KS, Yokum S (2013) Relative ability of fat and sugar tastes to activate reward, gustatory, and somatosensory regions. Am J Clin Nutr 98(6):1377–1384
13. Small DM, Jones-Gotman M, Dagher A (2003) Feeding-induced dopamine release in dorsal striatum correlates with meal pleasantness ratings in healthy human volunteers. Neuroimage 19(4):1709–1715
14. Tellez L, Medina S, Han W, Ferreira J, Licona-Limon P, Ren X et al (2013) A gut lipid messenger links excess dietary fat to dopamine deficiency. Science 341:800–802
15. Lutter M, Nestler EJ (2009) Homeostatic and hedonic signals interact in the regulation of food intake. J Nutr 139(3):629–632
16. Burger KS, Berner LA (in press) A functional neuroimaging review of obesity, appetitive hormones and ingestive behavior. Phys Behav. (available online)
17. Kelley AE, Baldo BA, Pratt WE, Will MJ (2005) Corticostriatal-hypothalamic circuitry and food motivation: integration of energy, action and reward. Physiol Behav 86(5):773–795
18. Berridge KC, Ho CY, Richard JM, DiFeliceantonio AG (2010) The tempted brain eats: pleasure and desire circuits in obesity and eating disorders. Brain Res 1350:43–64
19. Mahmoud S, Thorsell A, Sommer W, Heilig M, Holgate J, Bartlett S, Ruiz-Velasco V (2011) Pharmacological consequences of the A118G opioid receptor polymorphims on morphine- and fentanyl-mediated modulation of CA^{2+} channels in humanized mouse sensory neurons. Anesthesiology 115(5):1054–1062
20. Stice E, Spoor S, Bohon C, Veldhuizen MG, Small DM (2008b) Relation of reward from food intake and anticipated food intake to obesity: a functional magnetic resonance imaging study. J Abnorm Psychol 117(4):924–935
21. Bruce AS, Holsen LM, Chambers RJ, Martin LE, Brooks WM, Zarcone JR, Savage CR (2010) Obese children show hyperactivation to food pictures in brain networks linked to motivation, reward and cognitive control. Int J Obesity 34(10):1494–1500

22. Dimitropoulos A, Tkach J, Ho A, Kennedy J (2012) Greater corticolimbic activation to high-calorie food cues after eating in obese vs. normal-weight adults. Appetite 58(1):303–312
23. Frankort A, Roefs A, Siep N, Roebroeck A, Havermans R, Jansen A (2011) Reward activity in satiated overweight women is decreased during unbiased viewing but increased when imaging taste: an event-related fMRI study. Int J Obes 36(5):1–11
24. Holsen L, Savage C, Martin L, Bruce A, Lepping R, Ko E, Brooks W, Goldstein JM (2012) Importance of reward and prefrontal circuitry in hunger and satiety: Prader-Willi syndrome vs simple obesity. Int J Obes 36(5):638–647
25. Martin LE, Holsen LM, Chambers RJ, Bruce AS, Brooks WM, Zarcone JR, Savage CR (2010) Neural mechanisms associated with food motivation in obese and healthy weight adults. Obesity 18(2):254–260
26. Rothemund Y, Preuschhof C, Bohner G, Bauknecht HC, Klingebiel R, Flor H, Klapp BF (2007) Differential activation of the dorsal striatum by high-calorie visual food stimuli in obese individuals. Neuroimage 37(2):410–421
27. Stice E, Yokum S, Bohon C, Marti N, Smolen A (2010) Reward circuitry responsivity to food predicts future increases in body mass: moderating effects of DRD2 and DRD4. Neuroimage 50(4):1618–1625
28. Stoeckel LE, Weller RE, Cook EW 3rd Twieg DB, Knowlton RC, Cox JE (2008) Widespread reward-system activation in obese women in response to pictures of high-calorie foods. Neuroimage 41(2):636–647
29. Ng J, Stice E, Yokum S, Bohon C (2011) An fMRI study of obesity, food reward, and perceived caloric density. Does a low-fat label make food less appealing? Appetite 57(1):65–72
30. Lawrence NS, Hinton EC, Parkinson JA, Lawrence AD (2012) Nucleus accumbens response to food cues predicts subsequent snack consumption in women and increased body mass index in those with reduced self-control. Neuroimage 63(1):415–422
31. Nolan-Poupart S, Veldhuizen MG, Geha P, Small DM (2013) Midbrain response to milkshake correlates with ad libitum milkshake intake in the absence of hunger. Appetite 60(1):168–174
32. Burger KS, Stice E (2013) Elevated energy intake is correlated with hyperresponsivity in attentional, gustatory, and reward brain regions while anticipating palatable food receipt. Am J Clin Nutr 97(6):1188–1194
33. Schultz W, Dayan P, Montague PR (1997) A neural substrate of prediction and reward. Science 275(5306):1593–1599
34. Tindell AJ, Berridge KC, Aldridge JW (2004) Ventral pallidal representation of pavlovian cues and reward: population and rate codes. J Neurosci 24(5):1058–1069
35. Tobler PN, Fiorillo CD, Schultz W (2005) Adaptive coding of reward value by dopamine neurons. Science 307(5715):1642–1645
36. Balleine BW, Daw ND, O'Doherty JP (2008) Multiple forms of value learning and the function of dopamine. In: Glimcher PW, Fehr E, Camerer C, Poldrack RA (eds) Neuroeconomics: decision making and the brain. Academic, New York, pp 367–385
37. Robinson TE, Berridge KC (1993) The neural basis of drug craving: an incentive-sensitization theory of addiction. Brain Res Brain Res Rev 18(3):247–291
38. Smith K, Tindell A, Aldridge J, Berridge K (2009) Ventral pallidum roles in reward and motivation. Behav Brain Res 196(2):155–167
39. Flagel SB, Clark JJ, Robinson TE, Mayo L, Czuj A, Willuhn I, Akil H (2011) A selective role for dopamine in stimulus-reward learning. Nature 469(7328):53–57
40. Chouinard-Decorte F, Felsted J, Small DM (2010) Increased amygdala response and decreased influence of internal state on amygdala response to food in overweight compared to healthy weight individuals. Appetite 54:639
41. Demos KE, Heatherton TF, Kelley WM (2012) Individual differences in nucleus accumbens activity to food and sexual images predict weight gain and sexual behavior. J Neurosci 32(16):5549–5552
42. Sun X, Veldhuizen M, Wray A, De Araujo I, Small D (2013) Amygdala response to food cues in the absence of hunger predicts weight change. Appetite 60(1):168–174
43. Yokum S, Ng J, Stice E (2011) Attentional bias to food images associated with elevated weight and future weight gain: an fMRI study. Obesity (Silver Spring) 19(9):1775–1783

44. Yokum S, Gearhardt A, Harris J, Brownell K, Stice, E (in press). Individual differences in striatum activity to food commercials predicts weight gain in adolescence. Obesity
45. Geha PY, Aschenbrenner K, Felsted J, O'Malley SS, Small DM (2013) Altered hypothalamic response to food in smokers. Am J Clin Nutr 97(1):15–22
46. Stice E, Spoor S, Bohon C, Small DM (2008) Relation between obesity and blunted striatal response to food is moderated by TaqIA A1 allele. Science 322(5900):449–452
47. Nikolova YS, Ferrell RE, Manuck SB, Hariri AR (2011) Multilocus genetic profile for dopamine signaling predicts ventral striatum reactivity. Neuropsychopharmacology 36(9):1940–1947
48. Stice E, Yokum S, Burger KS, Epstein L, Smolen A (2012) Multilocus genetic composite reflecting dopamine signaling capacity predicts reward circuitry responsivity. J Neurosci 32(29):10093–10100
49. Jonsson EG, Nothen MM, Grunhage F, Farde L, Nakashima Y, Propping P, Sedvall GC (1999) Polymorphisms in the dopamine D2 receptor gene and their relationships to striatal dopamine receptor density of healthy volunteers. Mol Psychiatry 4(3):290–296
50. Asghari V, Sanyal S, Buchwaldt S, Paterson A, Jovanovic V, Van Tol HH (1995) Modulation of intracellular cyclic AMP levels by different human dopamine D4 receptor variants. J Neurochem 65(3):1157–1165
51. Seeger G, Schloss P, Schmidt MH (2001) Marker gene polymorphisms in hyperkinetic disorder-predictors of clinical response to treatment with methylphenidate? Neurosci Lett 313(1–2):45–48
52. Heinz A, Goldman D, Jones DW, Palmour R, Hommer D, Gorey JG, Weinberger DR (2000) Genotype influences in vivo dopamine transporter availability in human striatum. Neuropsychopharmacology 22(2):133–139
53. van Dyck CH Malison RT Jacobsen LK Seibyl JP Staley JK Laruelle M Gelernter J (2005) Increased dopamine transporter availability associated with the 9-repeat allele of the SLC6A3 gene. J Nucl Med 46(5):745–751
54. Lachman HM, Papolos DF, Saito T, Yu YM, Szumlanski CL, Weinshilboum RM (1996) Human catechol-O-methyltransferase pharmacogenetics: description of a functional polymorphism and its potential application to neuropsychiatric disorders. Pharmacogenetics 6(3):243–250
55. Yokum S, Marti N, Smolen A, Stice E (submitted) Multilocus genetic composite reflecting elevated dopamine signaling predicts future weight gain.
56. Wang GJ, Volkow ND, Fowler JS (2002) The role of dopamine in motivation for food in humans: implications for obesity. Expert Opin Ther Targets 6(5):601–609
57. de Weijer BA van de Giessen E van Amelsvoort TA Boot E Braak B Janssen IM Booij J (2011) Lower striatal dopamine D2/3 receptor availability in obese compared with non-obese subjects. EJNMMI Res 1(1):37
58. Kessler RM, Zald DH, Ansari MS, Li R, Cowan RL (in press) Changes in dopamine release and dopamine D2/3 receptor levels with the development of mild obesity. Synapse
59. Volkow ND, Wang GJ, Telang F, Fowler JS, Thanos PK, Logan J, Pradhan K (2008) Low dopamine striatal D2 receptors are associated with prefrontal metabolism in obese subjects: possible contributing factors. Neuroimage 42(4):1537–1543
60. Eisenstein SA, Antenor-Dorsey JAV, Gredysa DM, Koller JM, Bihum EC, Ranck SA, Hershey T (2013) A comparison of D2 receptor specific binding in obese and normal-weight individuals using PET with (N-[^{11}C]methyl)benperidol. Synapse 67(11):748–756
61. Haltia LT, Rinne JO, Merisaari H, Maguire RP, Savontaus E, Helin S, Kaasinen V (2007) Effects of intravenous glucose on dopaminergic function in the human brain in vivo. Synapse 61(9):748–756
62. Wilcox C, Braskie M, Kluth J, Jagust W (2010) Overeating behavior and striatal dopamine with 6-[F]-Fluoro-L-m-Tyrosine PET. J Obesity 2010:909348
63. Babbs RK, Sun X, Felsted J, Chouinard-Decorte F, Veldhuizen MG, Small DM (2013) Decreased caudate response to milkshake is associated with higher body mass index and greater impulsivity. Physiol Behav 121:103–111

64. Frank GK, Reynolds JR, Shott ME, Jappe L, Yang TT, Tregellas JR, O'Reilly RC (2012) Anorexia nervosa and obesity are associated with opposite brain reward response. Neuropsychopharmacology 37(9):2031–2046
65. Green E, Jacobson A, Haase L, Murphy C (2011) Reduced nucleus accumbens and caudate nucleus activation to a pleasant taste is associated with obesity in older adults. Brain Res 1386:109–117
66. Fetissov SO, Meguid MM, Sato T, Zhang LH (2002) Expression of dopaminergic receptors in the hypothalamus of lean and obese Zucker rats and food intake. Am J Physiol Regul Integr Comp Physiol 283(4):R905–R910
67. Geiger BM, Behr GG, Frank LE, Caldera-Siu AD, Beinfeld MC, Kokkotou EG, Pothos EN (2008) Evidence for defective mesolimbic dopamine exocytosis in obesity-prone rats. FASEB J 22(8):2740–2746
68. Huang XF, Zavitsanou K, Huang X, Yu Y, Wang H, Chen F, Deng C (2006) Dopamine transporter and D2 receptor binding densities in mice prone or resistant to chronic high fat diet-induced obesity. Behav Brain Res 175(2):415–419
69. Thanos PK, Michaelides M, Piyis YK, Wang GJ, Volkow ND (2008) Food restriction markedly increases dopamine D2 receptor (D2R) in a rat model of obesity as assessed with in-vivo muPET imaging ([11C] raclopride) and in-vitro ([3H] spiperone) autoradiography. Synapse 62(1):50–61
70. Stice E, Yokum S, Burger KS, Epstein LH, Small DM (2011) Youth at risk for obesity show greater activation of striatal and somatosensory regions to food. J Neurosci 31(12):4360–4366
71. Stice E, Yokum S, Blum K, Bohon C (2010) Weight gain is associated with reduced striatal response to palatable food. J Neurosci 30(39):13105–13109
72. Bello NT, Lucas LR, Hajnal A (2002) Repeated sucrose access influences dopamine D2 receptor density in the striatum. Neuroreport 13(12):1575–1578
73. Davis JF, Tracy AL, Schurdak JD, Tschop MH, Lipton JW, Clegg DJ, Benoit SC (2008) Exposure to elevated levels of dietary fat attenuates psychostimulant reward and mesolimbic dopamine turnover in the rat. Behav Neurosci 122(6):1257–1263
74. Geiger BM, Haburcak M, Avena NM, Moyer MC, Hoebel BG, Pothos EN (2009) Deficits of mesolimbic dopamine neurotransmission in rat dietary obesity. Neuroscience 159(4):1193–1199
75. Kelley AE, Will MJ, Steininger TL, Zhang M, Haber SN (2003) Restricted daily consumption of a highly palatable food (chocolate Ensure(R)) alters striatal enkephalin gene expression. Eur J Neurosci 18(9):2592–2598
76. Johnson PM, Kenny PJ (2010) Dopamine D2 receptors in addiction-like reward dysfunction and compulsive eating in obese rats. Nat Neurosci 13(5): 635–641.
77. Val-Laillet D, Layec S, Guerin S, Meurice P, Malbert CH (2011) Changes in brain activity after a diet-induced obesity. Obesity (Silver Spring) 19(4):749–756
78. Burger KS, Stice E (2012) Frequent ice cream consumption is associated with reduced striatal response to receipt of an ice cream-based milkshake. Am J Clin Nutr 95(4):810–817
79. Green E, Murphy C (2012) Altered processing of sweet taste in the brain of diet soda drinkers. Physiol Behav 107(4):560–567
80. Rudenga K, Small D (2012) Amygdala response to sucrose consumption is inversely related to artificial sweetener use. Appetite 58(2):504–507
81. Wise RA, Leone P, Rivest R, Leeb, K (1995 Elevations of nucleus accumbens dopamine and DOPAC levels during intravenous heroin self-administration. Synapse 21(2):140–148
82. Wise RA, Newton P, Leeb K, Burnette B, Justice JB (1995) Fluctuations in nucleus accumbens dopamine concentrations during intravenous cocaine self-administration in rats. Psychopharmacology 120(1):10–20
83. Ranaldi R, Pocock D, Zereik R, Wise A (1999) Dopamine fluctuations in the nucleus accumbens during maintenance, extinction, and reinstatement of intravenous d-amphetamine self-administration. J Neurosci 19(10):4102–4109

84. Hare TA, Camerer CF, Rangel A (2009) Self-control in decision-making involves modulation of the vmPFC valuation system. Science 324(5927):646–648
85. Hollmann M, Hellrung L, Pleger B, Schlogl H, Kabisch S, Stumvoll M, Horstmann A (2012) Neural correlates of the volitional regulation of the desire for food. Int J Obes 36(5):648–655
86. Kober H, Mende-Siedlecki P, Kross EF, Weber J, Mischel W, Hart CL, Ochsner KN (2010) Prefrontal-striatal pathway underlies cognitive regulation of craving. Proc Natl Acad Sci U S A 107(33):14811–14816
87. Siep N, Roefs A, Roebroeck A, Havermans R, Bonte M, Jansen A (2012) Fighting food temptations: the modulating effects of short-term cognitive reappraisal, suppression and up-regulation on mesocorticolimbic activity related to appetitive motivation. NeuroImage 60(1):213–220
88. Yokum S, Stice E (2013) Cognitive regulation of food craing: effects of three cognitive reappraisal strategies on neural response to palatable foods. Int J Obes 37(12):1565–1570
89. Stoeckel L, Ghosh S, Stern J, Keshavan A, Gabrieli J, Whitfield-Gabrieli, Evins A (2013) Real time fMRI neurofeedback: Effect on food and cigarette cue reactivity. Paper presented at the American College of Neuropsychopharmacology, Florida
90. Harris JL, Bargh JA, Brownell KD (2009) Priming effects of television food advertising on eating behavior. Health Psychol 28(4):404–413
91. Halford JC, Gillespie J, Brown V, Pontin EE, Dovey TM (2004) Effect of television advertisements for foods on food consumption in children. Appetite 42(2):221–225
92. Story M, French S (2004) Food advertising and marketing directed at children and adolescents in the US. Int J Behav Natr Phys Act 1(1):3
93. Ledikwe JH, Ello-Martin J, Pelkman CL, Birch LL, Mannino ML, Rolls BJ (2007) A reliable, valid questionnaire indicates that preference for dietary fat declines when following a reduced-fat diet. Appetite 49(1):74–83
94. Mattes RD (1993) Fat preference and adherence to a reduced-fat diet. Am J Clin Nutr 57(3):373–381
95. Stewart JE, Feinle-Bisset C, Golding M, Delahunty C, Clifton PM, Keast RS (2010) Oral sensitivity to fatty acids, food consumption and BMI in human subjects. Br J Nutr 104(1):145–152
96. Stice E, Yokum S, Burger K, Rohde P, Shaw H, Gau J (in press). A pilot randomized trial of a cognitive reappraisal obesity prevention program. Physiology and Behavior

Chapter 7
Molecular Mechanisms Involved in the Regulation of Food Intake

Emilio González-Jiménez

Introduction

Obesity is the most prevalent endocrine-metabolic disease in industrialized countries, and considered a serious public health problem worldwide [1]. Obesity is defined as excessive storage of energy as fat, a situation that leads to the development of a positive energy balance [2]. Among the factors that will shape the development of a positive energy balance, environmental and neuroendocrine factors highlights [3]. This complex regulatory process is controlled at the level of the central nervous system by the hypothalamus. In addition, the regulatory peptides involved with numerous synergistic or antagonistic actions are synthesized in different tissues, interacting together and with different neural signals driving the information to different cores. This in turn triggers a response in terms of initiating or completing food intake and therefore controls the increase or reduction of caloric expenditure [4].

Nutritional homeostasis (Fig. 7.1) comprehends the set of physiological processes involved in the mechanisms of digestion, absorption of nutrients, storing, and their use when appropriate [5]. All this with one goal: allow adequate growth in size and weight during childhood and adolescence, and in adulthood to acquire and maintain a proper weight [5]. Nutritional homeostasis process got its start with the ingestion of food and subsequent digestion and absorption of nutrients, in which participates numerous enzymes and hormones [6]. Simultaneously, there is a distribution for the reservoirs of glycogen in liver and muscle (deposits of immediate use) and triglycerides inside adipocytes, in the postprandial stage [7]. This cycle continues in the fasting phase. For the supply of nutrients, different metabolic processes (lipolysis, glycogenolysis and gluconeogenesis) are activated [8]. In relation to energy expenditure, it's a

E. González-Jiménez (✉)
Department of Nursing, Faculty of Health Sciences,
University of Granada, Avenida de la Ilustración s/n, CP 18016, Granada, (Spain)
e-mail: emigoji@ugr.es

C. Nóbrega, R. Rodriguez-López (eds.), *Molecular Mechanisms Underpinning the Development of Obesity*, DOI 10.1007/978-3-319-12766-8_7

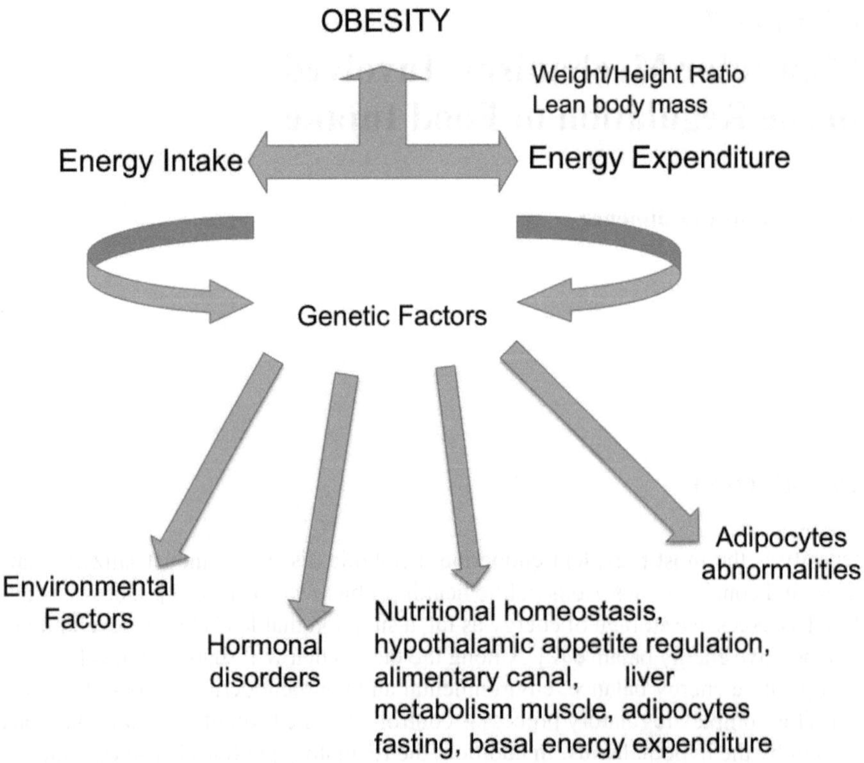

Fig. 7.1 Nutritional homeostasis. Major issues involved. (Adapted from Hermoso López [9])

sum between the spending in maintenance processes of the organism, the expenditure involved in physical activity, growth, and energy lost through secretions and body fluids. This complex process is interconnected and controlled by a large number of peripheral nerve signals, different molecules and mediators of neuroendocrine action integrated with the central nervous system (CNS) [10]. The structures responsible for controlling feelings of satiety and hunger in the healthy organism are located in the hypothalamus [10]. First, in the lateral nucleus, which is responsible for hunger. Second, in the ventromedial nucleus, which is responsible for management and control of satiety during food intake. These centers generate an efferent nerve response that controls the volume of energy stored or consumed in the short, medium and long term [11]. All aimed to keep the subject in a state of normal weight.

Mechanisms of Regulation of Short-Term Food Intake

Having appetite at a given time is closely related to the need of an immediate intake of nutrients. In this sense, the decrease in blood glucose and lipids is the main facilitator of hunger and thus an intake element, although it has been shown that this process is far more complex [12].

Food perception will involve different sensory signals (smell of food, taste, texture, temperature and even the appearance or presentation), which through the cranial nerves will be transmitted to the CNS [13]. Moreover, the start of insulin secretion is also mediated by these sensory signals, being thus another factor involved in activating nutrient intake. These set of signals is called the cephalic phase of feeding. To compensate these signals, there is a compensatory mechanism to decrease appetite [13]. This mechanism is possible due to oropharyngeal receptors, which control the total caloric intake ingested, generating signals to complete the ingestion.

If one or more daily food intakes are delayed, a decrease of blood glucose levels will happen. Consequently, the feeling of hunger will be stronger in order to regain the energy reserves consumed during the fasting period. This process is known as glucostatic theory [14]. The organic lipid reserves represent another source of information that the body uses to control the intake of nutrients. The central nervous system has receptors that regulate the concentrations of certain molecules as fatty acids, glycerol and 3-hydroxybutyrate, molecules that in turn provide information about the existing fat reserves in the body. This theory is known as lipostatic theory [15]. Despite differences between both theories, there is a consensus to accept that a decrease in the availability of nutrients correlates with feeding behavior. This theory is known as energostatic theory [16].

At gastrointestinal level, it has been identified several peptide molecules involved in inducing satiety. Its synthesis and secretion will be in proportion to the amount of ingested food [17], although their mechanisms of action will vary. They can act locally or be released into the bloodstream [18]. The information provided reaches the brain, specifically through vagal afferent fibers of peripheral nerves. From this region, the information will be transmitted in turn to the hypothalamus. Gastric distension achieved during intake will also enhance the effect of the peptide action already mentioned. The most relevant peptides are cholecystokinin (CCK), bombesin, glucagon, enterostatin, pancreatic polypeptide and amylin, among others. The satiating effect of cholecystokinin (CCK) was first described by Gibbs et al. [19] in 1973, who observed how the administration of CCK via peritoneal suppressed food intake in rats, and then in humans. Their release occurs by enteroendocrine cells of duodenal-jejunal mucosa in response to the intake of fats (unsaturated fatty acids, long chain mainly) and proteins [20]. Two types of receptors will mediate their action. The CCK-1 receptor type, formerly called CCK-A (alimentary), given its location in the alimentary tract, from where they stimulate nerve transmission by sending a message of satiety to the nucleus of the solitary tract [21]. These receptors are present in the pancreas, gallbladder, and pylorus and also in multiple locations of the CNS (nucleus of the solitary tract, area postrema and dorsomedial hypothalamus) [21]. The CCK-2 receptor type, previously known as CCK-B (Brain), which act in the brain are released in order to achieve an anorectic effect. These receptors are located in the vagus nerve and in the CNS [22]. Other peptides have different action mechanisms; bombesin reduces the volume of food eaten in each meal. Glucagon, through a mechanism of vagal action suppresses food intake but acting particularly on protein [23]. Enterostatin on the other hand, caused by fragmentation of the pancreatic lipase in the intestine, leads to a decrease in the duration of meal [23]. The pancreatic polypeptide and amylin by its interaction with the area postrema of the brain also induce a reduction in the food intake [23].

Mechanisms of Regulation of Medium-Term Food Intake

Appetite regulation in the medium term is regulated by peptide YY [24]. It is a hormone composed of 36 amino acids within the same group of pancreatic polypeptide and neuropeptide Y (NPY) [24]. It is synthesized in the L cells of the intestine, located in the more distal region of the same, in the pancreas and brain. Currently, it has been identified two types of Y endogenous peptides, subtype 1–36 and 3–36, the first being the more active metabolically. It was found that peptide YY 3–36 subtype specifically, has the ability to decrease by 36 % the volume of food ingested immediately and by 33 % the total volume within 24 h. This is why it is considered as the main regulator of food intake at intermediate periods of time [25].

Mechanisms of Regulation of Long-Term Food Intake

This type of long-term regulation occurs at the expense of peripheral adiposity signals and central neurotransmitters [26].

Peripheral Adiposity Signals with Anorectic Effect (Catabolic)

Two hormone molecules, the leptin and insulin compose the peripheral adiposity signals. Both of them act by inhibiting anabolic processes and stimulating catabolic metabolism [27].

Leptin

Leptin is the final product of ob-gene. This gene in humans is located in 7q31.3 chromosome 7q. It has 650 kb and consists of three separate exons, spaced by two introns. Exons two and three contain the region encoding leptin [28]. It consists of 167 amino acids whose sequence is very conserved across different species. For example, the mouse leptin maintains an 84 % homology to human leptin, and the rat around 83 % [28].

Through this hormone, hypothalamus controls the body nutritional intake modulating the food intake and energy balance [29]. Leptin triggers the activation of catabolic effector system. This cause a reduction of adiposity through an inhibition of appetite (anorexigenic effect), stimulating energy expenditure and disabling anabolic effector system, which is designed to increase body adiposity (via increased appetite), favoring the process of lipolysis in the adipose tissue [30]. Its synthesis is carried out by adipocytes, but also in organs and tissues such as the pituitary, hypothalamus, skeletal muscle, placenta, gastric mucosa and breast epithelium [31]. Its concentration will vary in proportion to the volume of fat reserves in the organism, that is, the amount of leptin produced by adipocytes and circulating is proportional to the volume of fatty acids accumulated inside adipocytes [31]. Its secretion in the body follows circadian rhythm pulses every 45 minutes [32]. Thus, the concentration of this hormone increases throughout the day until reaching its highest level at midnight, and continue decreasing until the beginning of a new cycle [33]. There are differences in leptin concentration in blood plasma by gender, being higher in women [34]. Androgens appear to be responsible for its lower concentration in the plasma of men [34]. According to physiology, circulating leptin exerts its anorexigenic effects through two processes of action. One mechanism is peripherally restricted, for which is required the existence of specific receptors for this hormone at the level of peripheral tissues and organs such as lung, kidney, liver, skeletal muscle, testis, pancreatic islets, stomach, adipose tissue (white and brown) and hematopoietic cells [35]. To exert its action at central level, circulating leptin must cross the blood-brain barrier, by the binding to specific receptors present in certain areas of the hypothalamus [36]. In Fig. 7.2 is represented the control mechanisms (acute and chronic) of feeding behavior, relying on the action of leptin.

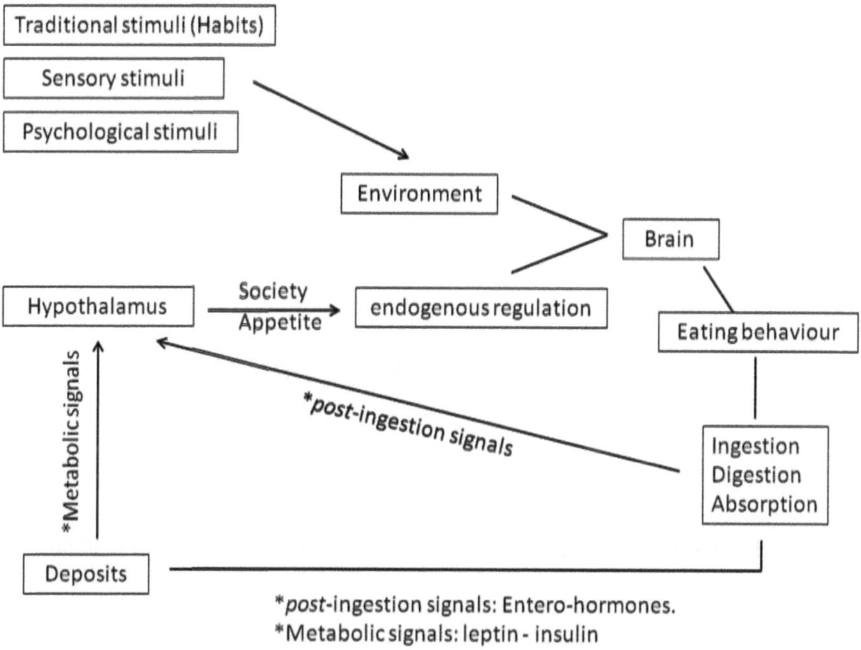

Fig. 7.2 Mechanisms involved in the regulation and appetite control. (Adapted from Arteaga [37])

The lack of leptin, by a shortfall in production or defects in specific receptors, will determine the development of early complications, such as severe early-onset obesity, hyperphagia, diabetes, and infertility [38]. It has been shown that exogenous inputs of leptin substantially improve these symptoms and their intensity [38]. However, in an almost paradoxical way, the majority of subjects with severe obesity (except those with a mutation in the ob gene) present increased levels of leptin and insulin in plasma [39]. This leads us to suspect a possible phenomenon of resistance to the action of these hormones, which is verified in the case of the insulin. Such leptin resistance may be mediated by process disturbances in the leptin transport through the endothelial cells of the blood-brain barrier. It could also be related with defects in the leptin receptor or by alterations in one or more of the neural systems that respond to leptin signaling [40, 41]. Although much is still unknown about leptin involvement in obesity, many aspects lead us to believe in obesity resulting from a leptin resistance mechanism rather than a deficiency in subjects with obesity [2].

Insulin

Insulin is also involved in the regulation of energy balance. The blood levels of the hormone are directly proportional to the body fat volume [42]. Circulating insulin, like leptin, can access the central nervous system, by a saturable transport process based in receptors present in endothelial cells of capillaries and cerebral vessels. Insulin receptors also are located in hypothalamic brain areas [42]. Despite the similarities between these molecules, the release of insulin, in contrast to leptin, is in response to the stimulation of a single meal. It has been proven, that administration of insulin in the central nervous system, causes considerable anorectic effect and a consequent loss of body weight, although it is true that this effects appear to be dose-dependent [43]. This fact has been verified by the reverse situation, i.e., lower insulin levels in the brain generate an increase of food intake and consequently a weight gain. This evidence, gives strength about the role of insulin as an anorectic agent. However, we must not forget other functions of this hormone and the effect that insulin exerts in the storing of ingested nutrients [44].

Peripheral and Central Adiposity Signals with Orexigenic Effect (Anabolic) and Anorexigenic (Catabolic)

In this section, we found a large group of molecules whose involvement and metabolic intervention will affect the production of an anabolic effect. From those, the peripheral ghrelin levels highlights [45], whereas at the central level, it stands out the neuropeptide Y (NPY), agouti-related protein (AgRP) and melanin-concentrating hormone (MCH) [46].

Peripheral Orexigenic Signals

Ghrelin

The isolation of an endogenous ligand for a predetermined orphan receptor whose activation induces a potent release of growth hormone (GH) was successfully confirmed in 1999 and received the nickname Ghrelin [47]. Ghrelin belongs to the family of secretagogues hormones (GH). It consists of 28 amino acids and its most active molecular form is acetylated by octanoic acid at its third serine residue [47]. The hormone structure determines its ability to cross the blood-brain barrier, relying in specific receptors in the hypothalamic arcuate nucleus, which are related to the eating behavior [48]. Its origin and production is outside the central nervous system, particularly at the level of the stomach, being the major hormone at the gastrointestinal level with orexigenic effect. However, this substance is also produced in other locations such as the hypothalamus, pituitary, pancreas, kidney, intestine, and heart cells. It has also been isolated from placenta and testis [49]. Regarding its release, higher hormone levels correspond to periods of fasting or states of cachexia, in which a feeling of hunger is generated, with low concentrations being found in obese and healthy subjects after intake of nutrients. This molecule has been shown to be very important in the regulation of food intake, through several mechanisms like its competitive nature with leptin or its interaction with the vagus nerve [50]. It this last mechanism, can induce neuronal activation in the nucleus of the solitary tract causing motility, gastric secretion and ultimately induction of appetite and food intake [51]. Nevertheless, the main course of its action takes place in the arcuate nucleus, exercising a stimulus for the synthesis of other orexigenic peptides such as NPY and AgRP [52].

Central Orexigenic Signals

Neuropeptide Y (NPY)

Neuropeptide Y is an important molecule with an orexigenic potential. It is synthesized in the hypothalamus, particularly in the arcuate nucleus. Its action is exerted at the central level in the paraventricular nucleus, where it stimulates and enhances food intake and thus weight gain. Its levels rise before the depletion of body fat reserves, which occurs in situations of fasting or uncontrolled diabetes mellitus

[53]. Moreover, the synthesis of this peptide is inhibited by leptin and its metabolic anorectic effect [54]. It has been proven that NPY decreases the stimulus of the sympathetic nervous system towards the brown adipose tissue. This situation results in a decrease of energy expenditure and a consequent increase and development of white adipose tissue [55]. Experimental studies in mice, which have a deficit of specific receptors for this orexigenic peptide, showed a development of a moderately obese phenotype [56].

Agouti-related Protein (AgRP)

It is a peptide molecule, whose discovery in the human brain was accidental [56]. It synthesis occurs in the hypothalamus, specifically at the level of the arcuate nucleus and is distributed to the paraventricular nucleus and other regions of the CNS [56]. Its relevance in the field of nutrition derives from The high potential as inducing agent of appetite and consequently weight gain. The mechanism of action is based on its antagonizing effect with respect to MC3 and MC4 receptors of the α-MSH [57]. Its action on the nervous system, unlike what happens with NY appears to be more prolonged in time achieving its effects over several days [58].

Melanin-concentrating Hormone (MCH)

The melanin-concentrating hormone (MCH) is another possible way or anabolic mechanism of action regarding body energy balance. This hormone is a neuropeptide of 19 amino acids whose synthesis is carried out in the lateral hypothalamus during periods of fasting and in response to a deficit of leptin levels [59]. Moreover, based on its inhibitory effect on the hypothalamus-pituitary-thyroid axis towards AgRP and NPY, we must consider that while controlling the degree of appetite it might also be involved in the reduction of energy expenditure [60]. In studies with mice, it has been shown that inactivation of the gene that produces this hormone generates hypophagia and extreme thickness [61]. Currently, numerous initiatives and prospects for therapy contemplating this route are being studied, as another possible key piece in this complex puzzle.

Central Anorectic Signals

A-Melanocyte Stimulating Hormone (α-MSH)

The melanocyte-stimulating hormone (α-MSH) is derived from the proteolysis of a prohormone called proopiomelanocortin (POMC). This process occurs in neurons of the arcuate nucleus and is activated by leptin. The anorectic action of this hormone is produced by binding to the MC4-R receptors, which represents its main agonist [62]. In mice, it has been verified that the inactivation of the gene encoding the MC4-R receptor yields obesity [63]. Studies with obese subjects demonstrate the existence of a large number of patients with mutations in the MC4 receptor, which are to some extent responsible for this hormone malfunction and subsequent energy imbalance in these individuals [64]. From this genetic alteration, generally emerge obesity and hyperphagia, hyperinsulinemia, and hyperglycemia [65]. Therefore, the presence of mutations in the MC4-R receptor are currently the main cause of monogenic obesity in humans [65].

Corticotropin Releasing Hormone (CRH) and Thyrotropin Releasing Hormone (TRH)

A level of the hypothalamus, and paraventricular nucleus neurons are synthesized two neuropeptides whose main effect appears to be the inhibition of appetite. These molecules are the corticotropin-releasing hormone (CRH) and thyrotropin-releasing hormone (TRH) [66]. The first increases energy expenditure at the expense of activation of the sympathetic nervous system. In addition exerts regulatory action in the pituitary-hypothalamus axis [67]. Similarly, TRH is involved in the regulation of this axis [67]. Both hormones were identified by their involvement in mediating the effects of leptin, as it has been shown that the synthesis of these hormones in the hypothalamus increases in response to leptin [68]. In Table 7.1, is summarized the hormones and peptides discussed above (and others), specifying its site of synthesis in the human body, together with its effects on food intake, body weight and energy expenditure.

Conclusions

Numerous studies have demonstrated the pathophysiologic complexity of the regulation of food intake and energy balance. Investigations confirm that endocrine signals originated in the gastrointestinal tract are critical in regulating the food intake and energy homeostasis. From those, with exception of ghrelin, the other gastrointestinal hormones have inhibitory effects of appetite. Moreover, the molecules involved in the central regulation of food intake are also important, as have been successfully implicated in obese phenotype. In this regard, a better understanding of the physiological processes by which these molecules regulates food intake will be crucial for the development of new strategies against eating disorders such as obesity.

Table 7.1 Biomolecules regulatory intake, body weight and energy expenditure. (Adapted from Solomon et al. [69])

Hormone/peptide	Intake	Body weight	Energy expenditure	Synthesis
AGRP	↑	↑	↓	Brain
Amylin	↓	↓?	↑?	Pancreas
Bombesin	↓	–	↑	Stomach, intestine and brain
CART	↓	↓	↑?	Brain
CCK	↓	–	–	Small intestine
Cytokines	↓	↓	–	Stomach and intestine
Corticosteroids	↑	–	–	Small intestine
CRH	↓	↓	↑	Brain
Endocannabinoids	↑	–	–	Small intestine
Galanin	↑	–	–	Small intestine
GALP	↓	↓	↑	Brain
Gastrin	↓			Stomach
Ghrelin	↑	↑	–	Stomach, intestine and brain
GIP	–?	↑	↓?	Intestine
GLP-1	↓	↓?	–	Stomach, intestine and brain
Glucagon	↓	↓	–	Pancreas
GRP	↓			Stomach
Insulin	↓	↓	↑	Pancreas
Leptin	↓	↓	↑	Adipose tissue and Stomach
MCH	↑	–	↓	Brain
Neurotensin	↓	–	–	Brain and stomach
NPY	↑	↑	↓	Brain
Obestatin	↓	↓	–	Stomach and intestine
Opioids	↑	–	–	Brain
Orexins	↑	↑	–	Brain
Oxyntomodulin	↓	–	–	Intestine and brain
Oxytocin	↓	–	–	Brain
Peptide YY	↓	–	–	Stomach
POMC	↓	↓	↑	Brain
PP	↓	–	–	Pancreas
TRH	↓	–	↑	Brain

↓ decrease, ↑ increase

References

1. Holmgren M, Lindgren A, de Munter J, Rasmussen F, Ahlström G (2014) Impacts of mobility disability and high and increasing body mass index on health-related quality of life and participation in society: a population-based cohort study from Sweden. BMC Public Health 14(1):381
2. González Jiménez E, Aguilar Cordero MJ, García García CJ, García López PA, Álvarez Ferre J, Padilla López CA (2010) Leptina: un péptido con potencial terapéutico en sujetos obesos. Endocrinol Nutr 57(7):322–327
3. Higgins JA (2014) Resistant starch and energy balance: impact on weight loss and maintenance. Crit Rev Food Sci Nutr 54(9):1158–1166
4. Liu X, Zhu Z, Kalyani M, Janik JM, Shi H (2014) Effects of energy status and diet on Bdnf expression in the ventromedial hypothalamus of male and female rats. Physiol Behav 130:99–107
5. Kirsz K, Zięba DA (2012) The discovery of neuromedin U and its pivotal role in the central regulation of energy homeostasis. Postepy Hig Med Dosw 16(66):196–203
6. Kirsz K, Zieba DA (2011) Ghrelin-mediated appetite regulation in the central nervous system. Peptides 32(11):2256–2264
7. Zieba DA, Kirsz K, Molik E, Romanowicz K, Wojtowicz AK (2011) Effects of orexigenic peptides and leptin on melatonin secretion during different photoperiods in seasonal breeding ewes: an in vitro study. Domest Anim Endocrinol 40(3):139–146
8. Zieba DA, Amstalden M, Williams GL (2005) Regulatory roles of leptin in reproduction and metabolism: a comparative review. Domest Anim Endocrinol 29(1):166–185
9. Hermoso López F (2002) Obesidad infantil. Concepto, valoración, clasificación, etiopatogenia y bases genéticas. In: Cañete Estrada R, Fernández García JM. (eds) Actualizaciones en Endocrinología Pediátrica.Ergon, Madrid, pp 81–94
10. Solomon A, De Fanti BA, Martínez JA (2004) Control del apetito y peso corporal: la ghrelina y la señalización orexigénica. Nutr Clín y Diet Hosp 55:13–27
11. Badman MK, Flier JS (2005) The gut and energy balance: visceral allies in the obesity wars. Science 307:1909–1914
12. Lénárd L, Karádi Z (2012) Regulatory processes of hunger motivated behavior. Acta Biol Hung 63(1):80–88
13. Granados K, Stephens BR, Malin SK, Zderic TW, Hamilton MT, Braun B (2012) Appetite regulation in response to sitting and energy imbalance. Appl Physiol Nutr Metab 37(2):323–333
14. Halford JC, Harrold JA (2012) Satiety-enhancing products for appetite control: science and regulation of functional foods for weight management. Proc Nutr Soc 71(2):350–362
15. Speakman JR, Stubbs RJ, Mercer JG (2002) Does body mass play a role in the regulation of food intake? Proc Nutr Soc 61(4):473–487
16. Friedam MI, Tordoff MG, Ramírez I (1986) Integrated metabolic control of food intake. Brain Res Bull 17:855–859
17. Neary NM, Goldstone AP, Bloom SR (2004) Appetite regulation: from the gut to the hypothalamus. Clin Endocrinol 60:153–160
18. Daly DM, Park SJ, Valinsky WC, Beyak MJ (2011) Impaired intestinal afferent nerve satiety signalling and vagal afferent excitability in diet induced obesity in the mouse. J Physiol 589(11):2857–2870
19. Gibbs J, Young RC, Smith GP (1973) Cholecystokinin decreases food intake in rats. J Comp Physiol Psychol 84:488–495
20. Fromentin G, Darcel N, Chaumontet C, Marsset-Baglieri A, Nadkarni N, Tomé D (2012) Peripheral and central mechanisms involved in the control of food intake by dietary amino acids and proteins. Nutr Res Rev 29:1–11
21. Wen D, Ma CL, Zhang YJ, Meng YX, Ni ZY, Li SJ, Cong B (2012) Cholecystokinin receptor-1 mediates the inhibitory effects of exogenous cholecystokinin octapeptide on cellular morphine dependence. BMC Neurosci 13(1):63

22. Laverman P, Joosten L, Eek A, Roosenburg S, Peitl PK, Maina T et al (2011) Comparative biodistribution of 12 ^{111}In-labelled gastrin/CCK2 receptor-targeting peptides. Eur J Nucl Med Mol Imaging 38(8):1410–1416
23. Nylec M, Olszanecka-Glinianowicz M (2010) A little-known new components of the appetite control. Postepy Hig Med Dosw 64:291–295
24. Simpson K, Parker J, Plumer J, Bloom S (2012) CCK, PYY and PP: the control of energy balance. Handb Exp Pharmacol 209:209–230
25. Suzuki K, Jayasena CN, Bloom SR (2011) The gut hormones in appetite regulation. J Obes 2011: 528 − 401
26. Moss C, Dhillo WS, Frost G, Hickson M (2012) Gastrointestinal hormones: the regulation of appetite and the anorexia of ageing. J Hum Nutr Diet 25(1):3–15
27. Marroquí L, González A, Neco P, Caballero-Garrido E, Vieira E, Ripoll C, Nadal A, Quesada I (2012) Role of leptin in the pancreatic β-cell: effects and signaling pathways. J Mol Endocrinol 49(1):9–17
28. Jéquier E (2002) Leptin signaling, adiposity, and energy balance. Ann N Y Acad Sci 967:379–388
29. Roubos EW, Dahmen M, Kozicz T, Xu L (2012) Leptin and the hypothalamo-pituitary-adrenal stress axis. Gen Comp Endocrinol 177(1):28–36
30. Robertson SA, Leinninger GM, Myers MG (2008) Molecular and neural mediators of leptin action. Physiol Behav 94:637–642
31. Karbowska J, Kochan Z (2012) Leptin as a mediator between obesity and cardiac dysfunction. Postepy Hig Med Dosw 66:267–274
32. Kul A, Baltaci AK, Mogulkoc R (2012) Effect of testosterone supplementation on leptin release in rats after castration and/or unilateral surrenalectomy. Endokrynol Pol 63(2):119–124
33. Kanoski SE (2012) Cognitive and neuronal systems underlying obesity. Physiol Behav 106(3):337–344
34. Theodore MD, Iosif MD, Sharon MD, Christos S, Mantzoros MD (2010) Narrative review: the role of Leptin in human physiology: emerging clinical applications. Ann Intern Med 152:93–100
35. Young AA (2012) Brainstem sensing of meal-related signals in energy homeostasis. Neuropharmacology 63(1):31–45
36. Tucholski K, Otto-Buczkowska E (2011) The role of leptin in the regulation of carbohydrate metabolism. Endokrynol Pol 62(3):258–262
37. Arteaga A (1997) Etiopatogenia de la obesidad. Boletín de la Escuela de Medicina 26(1), available in: http://escuela.med.puc.cl/publ/boletin/obesidad/EtiopatogeniaObesidad.html
38. Khan SM, Hamnvik OP, Brinkoetter M, Mantzoros CS (2012) Leptin as a modulator of neuroendocrine function in humans. Yonsei Med J 53(4):671–679
39. Kelly AS, Metzig AM, Schwarzenberg SJ, Norris AL, Fox CK, Steinberger J (2012) Hyperleptinemia and hypoadiponectinemia in extreme pediatric obesity. Metab Syndr Relat Disord 10(2):123–127
40. Jéquier E (2002) Leptin signalin, adiposity, and energy balance. Ann N Y Acad Sci 967:379–388
41. Schwartz MW, Woods SC, Porte D, Seeley RJ, Baskin DG (2000) Central nervous system control of food intake. Nature 404:661–671
42. Floyd ZE, Stephens JM (2012) Controlling a master switch of adipocyte development and insulin sensitivity: covalent modifications of PPARγ. Biochim Biophys Acta 1822(7):1090–1095
43. Diamant M (2007) Brain insulin signalling in the regulation of energy balance and peripheral metabolism. Ideggyogy Sz 60(3–4):97–108
44. Begg DP, Woods SC (2012) The central insulin system and energy balance. Handb Exp Pharmacol 209:111–129
45. Andrews ZB (2011) Central mechanisms involved in the orexigenic actions of ghrelin. Peptides 32(11):2248–2255
46. Parker JA, Bloom SR (2012) Hypothalamic neuropeptides and the regulation of appetite. Neuropharmacology 63(1):18–30

47. Kojima M, Hosoda H, Date Y, Nakazato M, Matsudo H, Kangawa K (1999) Ghrelin is a growth hormone-releasing acylated peptide from stomach. Nature 402:656–660
48. Cabral A, Suescun O, Zigman JM, Perello M (2012) Ghrelin indirectly activates hypophysiotropic CRF neurons in rodents. PLoS ONE 7(2):e31462
49. Cuomo R, D'Alessandro A, Andreozzi P, Vozzella L, Sarnelli G (2011) Gastrointestinal regulation of food intake: do gut motility, enteric nerves and entero-hormones play together? Minerva Endocrinol 36(4):281–293
50. Lazarczyk MA, Lazarczyk M, Grzela T (2003) Ghrelin: a recently discovered gut-brain peptide. Int J Mol Med 12:279–287
51. Inui A, Asakawa A, Bowers CY, Mantovani G, Laviano A, Meguid MM (2004) Ghrelin, appetite, and gastric motility: the emerging role of the stomach as an endocrine organ. FASEB J 18:439–456
52. Seoane LM, Tovar SA, Caminos JE, Nogueiras R, Diéguez C (2004) Ghrelina: un péptido regulador de la ingesta. Rev Esp Obes 2:31–42
53. Chen Z, Travers SP, Travers JB (2012) Activation of NPY receptors suppresses excitatory synaptic transmission in a taste-feeding network in the lower brain stem. Am J Physiol Regul Integr Comp Physiol 302(12):1401–1410
54. Zhaohui Z, Jingzhu Z, Guipeng D, Xuesong W, Yuanming Z, Yinping W, Yugui C (2012) Role of neuropeptide Y in regulating hypothalamus-pituitary-gonad axis in the rats treated with electro-acupuncture. Neuropeptides 46(3):133–139
55. Estemberg D, Sikora-Szubert A, Kowalska-Koprek U, Berner-Trabska M, Brzozowska M, Pasiński J, Swierczewski A, Karowicz-Bilińska A (2011) Body mass changes during pregnancy and concentration of insulin and neuropeptide Y in women with regard to the BMI. Ginekol Pol 82(12):892–899
56. Cone RD (1999) The central melanocortin system and energy homeostasis. Endocrinol Metab 19: 211 – 16
57. Pritchard LE, Armstrong D, Davies N, Oliver RL, Schmitz CA (2004) Agouti-related protein (83–132) is a competitive antagonist at the human melanocortin-4 receptor: no evidence for differential interactions with pro-opiomelanocortin-derived ligands. J Endocrinol 180:183–191
58. Fekete C, Wittmann G, Liposits Z, Lechan RM (2004) Origin of cocaine and anphetamine-regulated transcrip (cart)-inmunoreactive innervation of the hipothalamic paraventricular nucleus. J Comp Neurol 469:340–350
59. Szekely M, Petervari E, Balasko M (2010) Thermoregulation, energy balance, regulatory peptides: recent developments. Front Biosci (Schol Ed) 2:1009–1046
60. Nahon JL (2006) The melanocortins and melanin-concentrating hormone in the central regulation of feeding behavior and energy homeostasis. C R Biol 329(8):623–638
61. Wang Y, Ziogas DC, Biddinger S, Kokkotou E (2010) You deserve what you eat: lessons learned from the study of the melanin-concentrating hormone (MCH)-deficient mice. Gut 59(12):1625–1634
62. Reaux-Le Goazigo A, Bodineau L, De Mota N, Jeandel L, Chartrel N, Knauf C, Raad C, Valet P, Llorens-Cortes C (2011) Apelin and the proopiomelanocortin system: a new regulatory pathway of hypothalamic α-MSH release. Am J Physiol Endocrinol Metab 301(5):955–966
63. Peter JC, Zipfel G, Lecourt AC, Bekel A, Hofbauer KG (2010) Antibodies raised against different extracellular loops of the melanocortin-3 receptor affect energy balance and autonomic function in rats. J Recept Signal Transduct Res 30(6):444–453
64. Vaisse C, Clement K, Durand E, Hercber S, Guy-Grand B, Froguel P (2000) Melanocortin-4 receptor mutations are a frequent and heterogeneous cause of morbid obesity. J Clin Invest 106:253–262
65. Shintani M, Ogawa Y, Nakao K (2002) Obesity induced by abnormality in leptin receptor and melanocortin-4 receptor. Nihon Rinsho 60(2):404–409
66. Valassi E, Scacchi M, Cavagnini F (2008) Neuroendocrine control of food intake. Nutr Metab Cardiovasc Dis 18(2):158–168

67. Yada T, Kohno D, Maejima Y, Sedbazar U, Arai T, Toriya M, Maekawa F, Kurita H, Niijima A, Yakabi K (2012) Neurohormones, rikkunshito and hypothalamic neurons interactively control appetite and anorexia. Curr Pharm Des 18(31):4854–4864
68. Okada R, Kobayashi T, Yamamoto K, Nakakura T, Tanaka S, Vaudry H, Kikuyama S (2009) Neuroendocrine regulation of thyroid-stimulating hormone secretion in amphibians. Ann N Y Acad Sci 1163:262–270
69. Solomon A, Martínez JA (2006) Participación del sistema nervioso y del tracto gastrointestinal en la homeostasis energética. Rev Med Univ Navarra 50(1):27–37

Chapter 8
Epigenetics of Human Obesity: A Link Between Genetics and Nutrition

David Albuquerque, Licínio Manco and Clévio Nóbrega

Introduction

The prevalence of obesity has reached epidemic proportion becoming one of the world's greatest public health challenges, contributing to the increased risk of many chronic diseases, such as type 2 diabetes mellitus, hypertension, cardiovascular diseases and other co-morbidities [1]. Obesity is a multifactorial disorder resulting from interaction between lifestyle factors (such as dietary habits, sedentary behavior and other environmental exposures) and genetics. The excess of body fat was described as the consequence of successive positive energy balance, where the amount of intake energy is greater than the amount of energy expended [2]. It is evident that the environment as diet has a significant role in the development of obesity. However, several twin and adoption studies found a strong evidence for a genetic component to the risk of obesity estimated to account for 40–70% of body mass index (BMI) variation in children and adults [3–7].

The advent of Genome-Wide Associations Studies (GWAS) over the last 7 years appeared as a powerful approach to identify genetic variants associated with common diseases. Until now, GWAS found more than 35 genetic *loci* associated with BMI [8]. The most relevant GWAS of obesity phenotype identified a common single-nucleotide polymorphism associated with BMI located at the first intron of the *fat mass and obesity associated* gene (*FTO*) in 2007 [9]. Each A allele of the *FTO* rs9939609 polymorphism leads to an increases body weight by 1.5 kg in adults.

D. Albuquerque (✉) · L. Manco
Faculdade de Ciências e Tecnologia, University of Coimbra, Coimbra, Portugal
e-mail: dav.albuquerque@gmail.com

C. Nóbrega
Center for Neurosciences and Cell Biology, University of Coimbra,
3004-517 Coimbra, Portugal
e-mail: clevio@cnc.uc.pt

C. Nobrega, R. Rodriguez-López (eds.), *Molecular Mechanisms Underpinning the Development of Obesity*, DOI 10.1007/978-3-319-12766-8_8

However, all the obesity-associated common single nucleotide polymorphisms signals known to date have a modest effect explaining only 2–4 % of the estimated 40–70 % inherited fraction of genetic susceptibility to obesity [8,10]. Thus, much of the obesity-associated genetic variance remains unexplained. However, other different mechanisms could lead to inter-individual differences in complex traits.

The interaction between the environment and genes largely favor the obese state [11]. The epigenetic regulation of gene expression, which has emerged in the last few years as a new and promising field in science, could also play an important role in the etiology of obesity. Thus, epigenetic mechanisms could help to explain why common genetic variation cannot be expected to fully account for the measured heritability, while a predisposition to obesity can be attributed to the interaction between ancestral genetic and epigenetic factors.

What is Epigenetic?

Conrad Waddington formulated the concept of epigenetic landscape for the first time in 1942. He used this term to define a new branch of biology, studying the relations of cause and effect between genes and their products to produce the phenotype [12]. The term epigenetics comes from the Greek word epigenesis, referring to embryology and genetics "as the genotype gives rise to the phenotype during development" [13]. Epigenetics could be seen as a new field of genetics or the "new genetics" because many biological processes are controlled not through gene mutations, but rather through reversible and heritable epigenetics phenomena ranging from DNA methylation changes, histone modifications, and regulatory action of micro-ribonucleic acid (miRNAs) [14–16]. We can found extreme definitions of epigenetics in the literature. To Conrad Waddington, epigenetics was the study of epigenesis, which is how genotypes give rise to phenotypes during development [12]. Robin Holliday defined epigenetics as "the study of the mechanisms of temporal and spatial control of gene activity during the development of complex organisms" [17]. For Riggs et al. [18] epigenetics is "the study of mitotically and/or meiotically heritable changes in gene function that cannot be explained by changes in DNA sequence". Currently, we can define epigenetic as heritable changes that are mitotically stable (and potentially meiotically) and affect gene function, but do not involve changes in the DNA sequence [19]. They occur in diverse organisms and control several biological functions.

The experiences of an individual can "modify" genes by epigenetic marks which remain over the lifespan [20]. When we say that genes can be modified, it does not mean a change in the nucleotide sequence of DNA by mutation. Using a simple, however good metaphor, we can say that genetics refers to genes' "writing", and epigenetics to genes "reading". One of the best natural examples of influence in the epigenetic status is given by honey bee (*Apis mellifera*). Fertile queen and sterile workers are genetically identical at the larvae level. So, which makes a simple bee becomes a queen? Well, it is the fact that she fed exclusively of pure royal jelly, which will make her progress to the phenotypic characteristics of a queen [21].

On the other hand, the other bees fed on normal diet and become simple workers. On a molecular level, it was shown that royal jelly silences the expression of DNA methyltransferase (Dnmt3), a key driver of epigenetic reprogramming which codes for the enzyme that catalyze the transfer of methyl groups to DNA, suggesting that honeybees use DNA methylation to control the levels of activity of their genes [21, 22]. When bee larvae fed royal jelly, turns Dnmt3 "*off*" and certain genes jump into action turning into a queen with fully developed ovaries. On the contrary, when a bee larvae is not fed with royal jelly, Dnmt3 is active and the larval development follows into morphological and physiological features to the default worker variety. This example is simple but can help to understand how environment, in this specific case diet, induces epigenetic changes that lead to different developments from the same DNA genome. DNA methylation in honey bee is used for storing epigenetic information, and the use of this information can be differently altered by nutrition [21]. Developmental and reproductive transitions brought by environmental cue can also be observed in plants. Vernalization is a natural process of plants in temperate climates. This process consists of flowering early after exposed to the cold temperatures of winter. Vernalization involves epigenetic modifications by environment signals which instruct plants to flower when weather conditions are favorable [23].

One of the best-studied examples in mammals on how early environmental exposures interact with epigenetic gene regulation to influence the phenotype is probably the *Agouti* mouse viable yellow (A^{vy}) model [24, 25]. The murine *agouti* gene, influenced by the extent of DNA methylation established early in development, affects coat color that correlates with adult body weight. In brown fur and slim healthy mice, the agouti genes are kept in the "*off*" position by attaching methyl groups to prevent transcription. In yellow fur and obese mice, however, the same genes are not methylated and thus, are expressed or turned "*on*". Varying the mother's diet tend to produce offspring with a wide variation in individual coat color and obese phenotype because epigenetic modifications of *agouti* genes were established in early development [26, 27]. The variation in phenotypes is caused by DNA methylation patterns that were acquired during early embryonic development and passed through the female germline that results in stable intergenerational transmission [19, 28–30].

Thus, epigenetics refers to all cases where a gene sequence alone is not enough to describe what is happening in terms of phenotypic changes. A same gene could be read differently according to specific circumstances. The DNA sequence is not modified, but the proteins codified from it could be produced in different moments or in different localizations, following epigenetics marks present on these genes. We have to remember that each cell that makes up the human body is composed by approximately 20,000–25,000 genes, all of them containing the same DNA sequence. So, epigenetics can be thought as a process that regulates gene expression in a given cell, leading to its cellular phenotype. Different gene expression results from an active gene "*turned off*" or "*turned on*" to express themselves. The gene environment is composed by several other components, starting with the cell including the nucleus (where genes are located), cytoplasm and organelles such are mitochondria, but also by tissues, organs, and the set of life conditions suffered by the individual (alimentation, physical activity). All these conditions are susceptible to modify the expression of our genes.

Epigenetic Mechanisms

The most extensively studied epigenetic mechanisms in mammalians are described in Fig. 8.1 and comprise DNA methylation, histone modifications and non-coding RNAs (e.g. microRNAs) [31–34]. All these mechanisms in combination with the underlying genetic sequence confer unique transcriptional instructions, and it is because of these mechanisms that a same genotype can generate different cells, lineages, organs and phenotypes.

Mechanisms of DNA Methylation

DNA methylation is a biologic process that consists in the addition of a methyl group to DNA cytosine residues (Fig. 8.1). The most common methylation is that of the carbon-5 position of cytosine (C) in the context of the CpG dinucleotides, a C

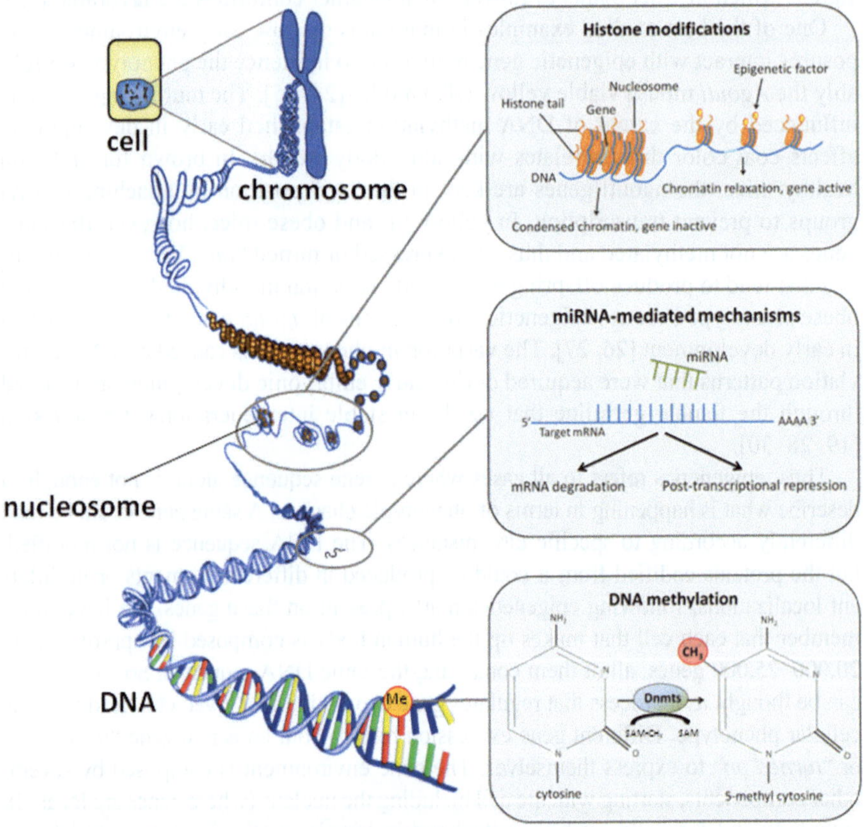

Fig. 8.1 Schematic representation showing how our DNA is packaged to fit inside cells, and the three fundamental mechanisms of epigenetic involved in gene expression regulation

followed by guanine (G) nucleotides linked by a phosphate group (p) which occurs mainly in gene promoter regions, and is usually associated with gene silencing [35, 36]. This stable epigenetic mark occurs in 70–80% of all CpG dinucleotides, and constitutes the most common modification in mammalians genomes that is transmitted through DNA replication and cell division [37]. It is also the well-studied of the epigenetic mechanisms which is involved in processes such as genomic imprinting and silencing of transposable elements [19]. The universal methyl donors are DNA methyltransferases (Dnmts) that catalyze the transfer of a methyl group from *S*-adenosylmethionine (SAM) to the carbon-5 of the cytosine pyrimidine ring, resulting in 5-methylcytosine (5mC) [38]. Four members of DNA methyltransferases have been identified in humans and mice involved in establishing and maintaining DNA methylation patterns (Dnmt1, Dnmt2, Dnmt3A and Dnmt3B) [39]. The Dnmt3A and Dnmt3B are responsible for establishing the initial *de novo* methylation pattern, whereas Dnmt1 ensures this pattern during the cell cycle. CpG dinucleotides are not randomly distributed throughout the genome but are clustered in regions known as CpG islands, that is, unmethylated GC-rich regions that contain high densities of CpG dinucleotides and are positioned in the promoter regions at the 5'ends of many genes [35, 40]. Hypermethylation of these CpG islands is associated with transcriptional repression, whereas hypomethylation is associated with transcriptional activation [40].

DNA methylation is very important in the establishment of the inactive X chromosome and in imprinting genes, as this mechanism is essential to maintain the silenced state [41]. Through the X-chromosome inactivation process, which has only been shown in somatic cells of female mammals with a placenta, only one of the two X-chromosomes is transcriptionally active whereas the other is inactive and its genes remain silent. Genomic imprinting is associated with gene expression whereby only one of the parental alleles is expressed.

Environmental perturbations during periods when DNA methylation patterns are induced (mainly during embryogenesis or in early postnatal life) can impair the program of gene silencing or activation with potential long-term adverse consequences. Thus, the measurement of DNA methylation levels in a genome-wide manner would be useful in studying the mechanisms of epigenetic control that are involved in gene regulation and imprinting.

Mechanisms of Histone Modifications

Each human diploid cell contains approximately two meters of chromosomal DNA packaged into the nucleus. This compaction is achieved by histones, which are small basic proteins consisting of a globular domain and a more flexible and charged NH_2-terminus (histone "tail") [42]. The result of this DNA-protein complex is called chromatin [43], and appeared similar to a "beads-on-a-string" [44] (Fig. 8.1). The basic unit of chromatin is the nucleosome "beads", which is structured by a paired of four histone (H) partners: H3 and H4 tetramer, flanked by two H2A-H2B dimmers come together to form a histone octamer, which binds and wraps about 1.7

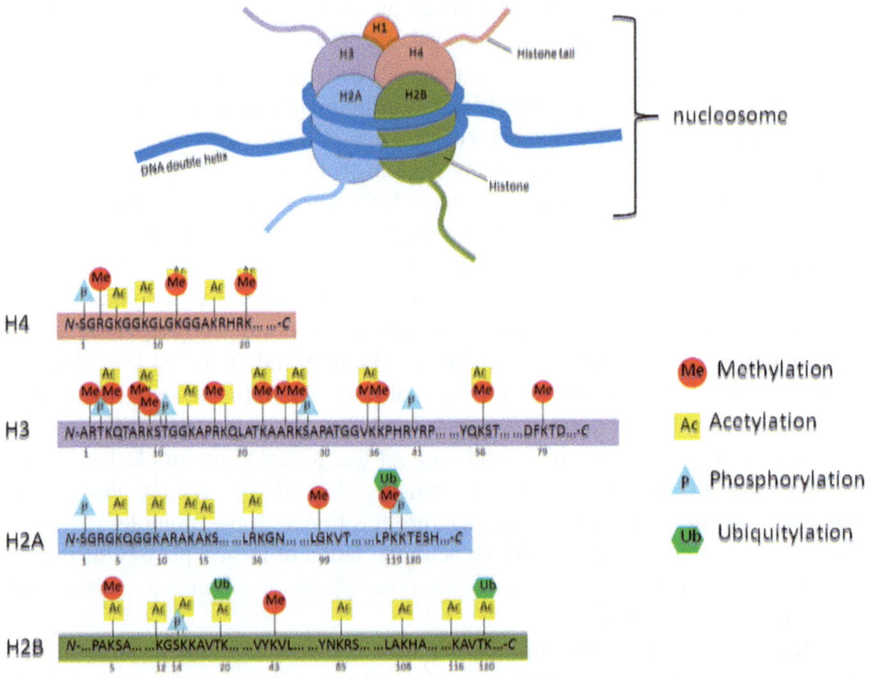

Fig. 8.2 The figure illustrates nucleosome models and major post-transcriptional modifications, which mainly occur in histone tails. These modifications play essential roles in genes expression regulation and disease processes. The main post-transcriptional modifications are depicted in this figure: methylation (*red*), acetylation (*yellow*), phosphorylation (*blue*) and ubiquitination (*green*). The number in *black* under each amino acid represents its position in the sequence

turns of DNA, or about 146 base pairs [45, 46] (Fig. 8.2). Another histone, called H1, wraps another 20 base pairs, resulting in two full turns around the octamer, stabilizing the nucleosome [47]. Histones have an active function in the regulation of chromatin structure and also in gene expression. There are a large variety of post-translational modifications that can alter histone tails (short arms that are separate from the main structure) such as acetylation, methylation, phosphorylation, ubiquitination [48], and sumoylation [49]. These modifications induce changes of the chromatin structure and subsequently affect gene expression (Fig. 8.2). Histone tail acetylation and histone methylation generally occurs on conserved lysine (K) on the N-terminal tails of all four-core histones. Histone acetylation facilitates "spacing out" of nucleosomes allowing easier access to transcription factors and is, in general, considered the most often active mark region of transcription [50, 51]. Methylation can be present in both active and inactive regions of chromatin [50]. Some evidences shows that deacetylation of histones together with methylation of H3K9 and of H3K27 are associated with silencing gene expression. In contrast, acetylation of histones and demethylation of H3K9 and H3K27, together with methylation of H3K4, is observed with active transcription of the associated gene [52]. Interactions between histone deacetylases (HDACs), histone methyltransferases and methyl-cytosine-binding proteins lead to the recruitment of DNA methyltransferases [50]. There could exist a possible "histone code" that characterizes all histone modifications that can be read and interpreted by different cellular factors [48].

Mechanisms of Non-Coding RNAs

Non-coding RNAs, including small miRNAs, also contribute to the regulation of the epigenome [31] (Fig. 8.1). These small molecules are endogenous short single-stranded non-protein coding gene products, constituted by approximately 19–25 nucleotides long [52, 53], which regulate gene expression at the post-transcriptional level and are conserved across species [54]. miRNAs act via base-pairing with complementary sequences within the 3' UTR of multiple target mRNA molecules, usually resulting in their silencing via translational repression or target degradation [55]. It has been estimated that the human genome encode more than 1000 miRNAs, which could regulate nearly 30 % of human genes [56, 57]. miRNAs were found to be involved in a variety of biological functions with a wide impact in cell proliferation, cell differentiation, including energy and fat metabolism [58]. Recently, several studies have shown that miRNAs can regulate the expression of the epigenetic machinery [59]. They have demonstrated that miRNAs acts as central modulators in adipogenesis [53]. Many miRNAs have a potential role in controlling adipocyte number and size, due to the regulation, enhancement, and inhibition of adipogenesis [60]. Normally, miRNAs are downregulated in obesity and upregulated during adipogenesis [53, 61]. Several studies showed that miRNAs expression in preadipocytes is altered during fat cell development and in obesity [60]. There are complex networks between miRNAs and epigenetic mechanisms to enhance the robustness of the human gene regulatory network. The understanding of the miRNAs regulatory roles in modulating energy balance, adipogenesis, and their contribution to obesity is needed to highlight our knowledge in obesity etiology.

None of these mechanisms operates separately. The epigenome, which comprises the overall epigenetic mechanisms, is working by sequence-specific transcription factors to coordinately regulate gene expression. The link between DNA methylation and histone modification have been largely demonstrated in gene silencing [62, 63]. These two epigenetic mechanisms are strongly connected one with each other in a complex interplay with the nucleosome structure. Methylation of cytosine within CpG islands is associated with binding proteins and subsequent recruitment of enzymes that catalyse histone modifications [33]. DNA methylation and miRNAs signal transcriptional silencing genes as specific histone modifications and their associated proteins may be characteristic of either gene silencing or gene expression [52].

Some "Epiobesogenic" Gene Studies

It is well known that epigenetic marks are tissue specific [64, 65]. The pathophysiology of obesity is associated with extensive gene expression changes in tissues throughout the body [51]. In this sense, it would be of great interest to study the baseline epigenetic differences in obesity-related tissues such as liver, muscle, pancreatic islets or hypothalamus [66, 67]. However, these tissues are largely inaccessible and, although it may not be representative of those alterations happening in more obesity-relevant tissues, peripheral blood white cells are being preferably used to explore the epigenetic characterization of humans at risk of obesity (Table 8.1). Relatively few studies have explored also other human tissues including umbilical cord blood, adipose cells or saliva. These studies have focused mostly on characterizing the methylation status of selected CpG sites in candidate genes, although the first Epigenome-Wide Association Studies (EWAS) are now emerging as large-scale studies to perform associations between methylation patterns at CpG sites and BMI. Moreover, there is a lack of independent replication studies in the field. In addition, to join controversy, there is no consensus to which cell type should be explored in EWAS. Further studies are needed in this field of research to corroborate this information. It is predictable that in the near future EWAS studies will arise identifying new pathways associated with obesity providing new data, which could help prevention.

Some Studies on DNA Methylation by Tissue Specificity

Blood cells are being preferably used to study epigenetic marks because of their less-invasive nature regarding sample collection, and several recent works evidenced methylation changes in blood leukocyte DNA associated with the susceptibility to obesity.

Regarding the most important human obesity related gene, hypermethylation within the first intron of the *FTO* gene using DNA from peripheral whole blood, has been associated with obesity and Type 2 Diabetes Mellitus [68–70]. Although the function of *FTO* remains unknown, it is known that it encodes a demethylase enzyme able to remove methyl groups from DNA residues [69, 71]. In the Almén et al. [69] study an association of the BMI risk allele rs9939609-A was found with the methylation status of several genes, including lysyl-tRNA synthetase (*KARS*), telomeric repeat binding factor 2, interacting protein (*TERF2IP*), dexamethasone-induced protein (*DEXI*), musashi RNA-binding protein 1 (*MSI1*), stonin 1 (*STON1*) and breast carcinoma amplified sequence 3 (*BCAS3*). However it remains unclear whether the association of *FTO* gene with obesity could be explained by epigenetic mechanisms.

Most recently, Xu et al. [72] compared the methylation profile in peripheral blood samples from 48 obese *versus* 48 lean African-American young (14–20 years-old), using 470,000 CpGs sites, and found that differentially variable CpG

Table 8.1 Studies with examples of human genes related to obesity through epigenetic mechanisms

Gene symbol/EWAS	Associated genes	Epigenetic mechanisms	Tissue	Study sample	Role in obesity	Reference
EWAS	*UBASH3A*, *TRIM3*	DNA methylation	Peripheral blood leukocytes	14 African-American men (14–18 years-old) Replication: 46 Obese and 46 lean African-American men	Obesity	[73]
KvDMR1, *H19*, *IGF2*, *GRB10*, *MEST*, *SNRPN*, *GNAS*	–	DNA methylation	Saliva	16 monozygotic twins discordant pair for BMI Caucasian (~ 30.9 year-old)	BMI	[84]
MCHR1	*MCHR1*	DNA methylation	Whole blood	49 Caucasian individuals (German) (21–78 years-old)	BMI	[132]
EWAS in individuals carriers *FTO* risk allele (rs9939609)	*KARS*, *TERF2IP*, *DEXI*, *MSI1*, *STON1*, *BCAS3*	DNA methylation	Whole blood	33 obese and 24 normal-weight preadolescent girls Caucasian (Greek) (9–13 years-old)	Obesity	[69]
CLOCK, *BMAL1* and *PER2*	*CLOCK*, *BMAL1* and *PER2*	DNA methylation	White blood cells	20 normal-weight, 20 overweight/obese and 20 morbidly obese women (Spanish) (25–53 years-old)	BMI, adiposity, weight loss, obesity	[133]
ALOX12, *ALPL*, *BCL2A1*, *CASP10*, *CAV1*, *CCL3*, *CD9*, *CDKN1C*, *DSC2*, *EPHA1*, *EV12A*, *HLA*, *IRF5*, *KRT1*, *LCN2*, *MLLT4*, *MMP9*, *MPL*, *NID1*, *NKX3*, *PMP22*, *S100A12*, *TAL1*, *VIM*	*ALPL*, *CASP10*, *CDKN1C*, *EPHA1*, *HLADOB3*, *IRF5*, *MMP9*, *MPL*, *NID1*	DNA methylation	Umbilical cord blood	178 Great-Britain children (~ 9.0 year-old)	BMI, fat mass, lean mass, height	[80]
SLC6A4	*SLC6A4*	DNA methylation	Peripheral blood leukocytes	84 monozygotic twin pairs Caucasian (~ 55.1 year-old)	BMI, body weight, waist circumference	[79]

Table 8.1 (continued)

Gene symbol/EWAS	Associated genes	Epigenetic mechanisms	Tissue	Study sample	Role in obesity	Reference
EWAS	*HSP90B3P*, *NAV1*, *NR5A2*, *CCDC48*, *GPR125*, *SNCA*, *EHMT2*, *IER3*, *SERPINB1*, *STX1A*, *PVT1*, *LHX6*, *ENKUR*, *CTTN*, *HCCA2*, *PKNOX2*, *ANO2*, *ITPR2*, *RB1*, *PACS2*, *CRTC3*, *KIFC3*, *MIR1910*, *ZFHX3*, *MSI2*, *RPTOR*, *TRPM4*, *C20orf160*, *LOC647979*, *MLC1*, *CDX4*, *KCND1*	DNA methylation, miRNA expression	Adipose tissue	31 healthy Caucasian men (Swedish) (~37.4 year-old)	Adipocyte metabolism	[86]
EWAS	Differences between number of differentially methylated CpG sites and number of differentially variable CpG sites	DNA methylation	Peripheral blood leukocytes	48 obese and 48 lean African-American (14–20 years-old)	Obesity	[72]
EWAS	*ZCCHC10*	DNA methylation	Umbilical cord blood	308 Black mother-infant pairs.	BMI	[134]
MEG3, *MEST*, *PEG3*, *PLAGL1*, *SGCE/PEG10*, *NNAT*	*MEST*, *PEG3*, *NNAT*, *PLAGL1*, *MEG3*	DNA methylation	Umbilical cord blood leukocytes	92 American newborns	Obesity	[83]

Table 8.1 (continued)

Gene symbol/EWAS	Associated genes	Epigenetic mechanisms	Tissue	Study sample	Role in obesity	Reference
365 miRNAs	25 miRNAs differently expressed in obese women and 7 obesity-specific miRNAs (miR-422b, miR-219, miR575, miR523, miR-579, miR-618 and miR-659)	miRNA expression	Whole blood and cord blood	10 obese 5 control Caucasian pregnant women (29–31 years-old)	Obesity	[135]
EWAS	*AQP9*, *DUSP22*, *HIPK3*, *TNNT1* and *TNNI3*	DNA methylation	Venous blood	24 (12–16 years-old) overweight/obese Spanish adolescents Replication: 83 overweight/obese Spanish adolescents	Weight, BMI-SDS, body fat	[136]
IGF2 and *H19*	*IGF2*, *H19*	DNA methylation	Umbilical cord blood leukocytes	79 newborns	Obesity	[82]
EWAS	5 meQTLs	DNA methylation, miRNA expression	Adipose tissue	19 obese and 19 non-obese men Caucasian (61.1 year-old) Replication: 181 female from Twins UK resource	BMI	[85]
RXRA, *eNOS*, *SOD1*, *IL8*, *PI3KCD*	*RXRA*, *eNOS*	DNA methylation	Umbilical cord tissue	78 Caucasian. women (≥16 years) Replication: 239 children	Fat mass and %fat mass	[81]
LEPR, *POMC*, *GHSR*, *NPY*	*POMC*, *NPY*	DNA methylation	Peripheral blood leukocytes	104 obese Caucasian Spanish	Body weight	[75]

sites were more variable in cases than controls. They also found that genes harbouring differentially methylated CpG sites and genes harbouring differentially variable CpG sites showed significant enrichment of genes identified by GWAS on obesity and obesity-related traits, supporting their roles in the etiology and pathogenesis of obesity in which either sequence variants or epigenetic variations may change gene functions.

Peripheral blood mononuclear cells were used in a study by Wang et al. [73] reporting CpG sites in ubiquitin associated and SH3 domain containing A (*UBA-SH3A*) and tripartite motif containing 3 (*TRIM3*) genes with different methylation profiles when comparing obese *versus* lean individuals.

The relations between an exposure of famine periods and DNA methylation have also been found in the Dutch Hunger Winter Families Study [20, 29] using DNA from whole blood. Individuals who were prenatally exposed to famine in 1944–1945 at the end of World War II showed, decades later, less DNA methylation in *loci* such as the imprinted insulin-like growth factor 2 (*IGF2*) and *INS-IGF2* Read-through (*INSIGF*), and increased DNA methylation for the genes *GNASAS*, *MEG3*, *IL10*, *ABCA1* and *LEP*, in parallel with impaired glucose tolerance compared with their unexposed same-sex siblings [20, 29, 74]. Regarding the exposure to an high-fat diet, studies revealed that it was not only associated with peripheral insulin resistance, but also influence DNA methylation of genes such the peroxisome proliferator-activated receptor gamma, coactivator 1-alpha (*PPARGC1A*) in skeletal muscle [75].

Peripheral blood leukocytes have also been used in DNA methylation analyses to predict the susceptibility to weight regain in appetite-related genes proopiomelanocortin (*POMC*), neuropeptide Y (*NPY*), which play a central role in the maintenance of energy homeostasis. In a recent study, Crujeira et al. [76] found that *NPY* and *POMC* methylation levels reflect a putative epigenetic regulation implicated in the weight regain process. Also, the promoter region hypermethylation of the serotonin transporter gene (*SLC6A4*), previously related with eating disorders and body weight [77, 78], was associated with an increase in BMI, body weight and waist circumference [79].

Umbilical cord blood had also been used to investigate epigenetic marks related to intrauterine environmental exposures [80]. Regarding obesity, several reports performed in cord blood samples evidenced epigenetic modifications linking early life exposures with later body phenotypes. In a recent report, Relton et al. [81] has evidenced that DNA methylation patterns in 9 of 24 (37.5%) genes at birth, show association with at least one index of body composition (BMI, fat mass, lean mass, height) at age 9 years. This observation suggests that variation in DNA methylation patterns at birth in multiple target genes may influence body size in childhood. Moreover, maternal diet can alters later child's adiposity, accompanied by epigenetic changes in genes controlling the energy homeostasis. Godfrey et al. [82] has evidenced that the methylation of CpGs from different gene promoters (*RXRA*, *eNOS*) at birth is associated with adiposity in later childhood. Furthermore, associations between pre-conceptional obesity and DNA methylation profiles in the offspring using DNA from umbilical cord blood leukocytes have also been observed

[83]. Parental pre-conceptional environmental exposures could have an effect in the health status of the offspring in later life. Pre-pregnancy maternal obesity was found to alter offspring DNA methylation in genes relevant to the development of complex disorders, including obesity, providing evidence of trans-generational influence on obesity susceptibility via epigenetic marks [8]. In two recent studies regarding parental obesity conducted by Soubry et al. [83, 84] it has been observed an association between DNA methylation profiles (especially differentially methylated regions) at human imprinted genes, such as mesoderm-specific transcript (*MEST*), paternally expressed gene 3 (*PEG3*), and neuronatin (*NNAT*), in children born from obese parents, when compared with children born from non-obese parents. Changes related to maternal obesity was also detected at *loci* pleiomorphic adenoma gene-like 1 (*PLAGL1*), maternally expressed gene 3 (*MEG3*) and Imprinted Maternally Expressed Transcript (Non-Protein Coding) (*H19*) [83, 84]. Hypomethylation at the *IGF2* gene was associated with paternal obesity [83]. These results points to a pre-conceptional influence of parental life-style or over-nutrition on the reprogramming of imprint marks during gametogenesis and early development [83, 84]. The trans-generational effects of parental obesity can influence the offspring's future health status. These reports have evidenced that peri-natal events are important in defining the epigenetic marks that will persist until the adult age and could be used as early prognostic markers to identify those individuals with more risk to develop later obesity. However, the knowledge of mechanisms by which maternal nutritional environment induces such changes remains in the beginning.

Saliva has also been used for DNA methylation analysis [80]. Using this tissue, Souren et al. [85] investigate in monozygotic twins discordant for BMI whether differential methylation at nine target differentially methylated regions most representative for abnormal growth syndromes contribute to the development of non-syndromic overweight; however no significant correlations between BMI differences and DNA methylation variability was found meaning that buccal cells DNA might not necessarily reflect the methylation status in whole blood or adipose tissue.

Adipose tissue represents an accessible and relevant tissue to investigate obesity-related traits. Generally, this tissue is collected from abdominal fat or subcutaneous biopsies. In a recent study, Drong et al. [86] investigated the profiling of 27,718 genomic regions using differential methylation hybridization in abdominal subcutaneous adipose tissue samples of 38 unrelated individuals, together with genotypes at 5,227,243 polymorphisms and expression of 17,209 mRNA transcripts, aiming to find genetic variation that influences methylation status in abdominal subcutaneous adipose tissue. They could not found any differential methylation hybridization probes that were significantly associated with BMI, but rather observed association between expression levels of two mRNA transcripts (*TNFRSF11B*, Tumor necrosis factor receptor super-family member 11B, and *GOT1*, Aspartate aminotransferase) and cis-methylation status, indicating that DNA CpG methylation in abdominal subcutaneous adipose tissue is partly under genetic control. In another study, Rönn et al. [87] performed a EWAS to evaluate the possible alteration of DNA methylation patterns after a 6 month exercise intervention. For this purpose 476,753 CpG sites were analyzed in 23 healthy men, with a previous low level of physical activity, and observed a global DNA methylation change in 17,975 CpG sites in response to physical activity.

There are several evidences suggesting that miRNAs play an essential role in regulating the pathways in adipose tissue differentiation [88]. Normally, miRNAs induced during adipogenesis are downregulated in obese subjects [53]. Several studies have been profiling the expression of many miRNAs during adipogenesis of human and rodent adipocytes. Xie et al. [89] profiled the expression of >370 miRNAs during adipogenesis of preadipocyte 3T3-L1 cells and adipocytes from leptin deficient ob/ob and diet-induced obese mice, observing that miRNAs that were induced during adipogenesis were downregulated in adipocytes from both types of obese mice and vice versa, and that ectopic expression miR-103 and miR-143 in preadipocytes accelerated adipogenesis. Expression levels of miR-143 were also found correlated with body weight and mesenteric fat weight [90]. The effect of miR-519d on adipogenesis was similar to that of miR-143 [57]. Klöting et al. [91] identified in abdominal subcutaneous and intra-abdominal omental adipose tissue some miRNAs significantly associated with fat depot at specific pattern. The expression of some of them was correlated with key metabolic parameters, including visceral fat area, fasting plasma glucose, and circulating leptin, adiponectin. Ortega et al. [92] found in obese subjects different expression profiles of miRNAs, up or down regulated, in fat cells and subcutaneous adipose tissues. These studies revealed that miRNAs might represent new biomarkers and potential based therapy targeting obesity and obesity-related disorders.

We are What We Eat: Nutritional Influences in Epigenetics

Of all the potential environmental factors affecting epigenetic mechanisms, nutrition appears to be the most influential, especially during early life, in the way that nutritional changes can modify patterns of gene expression influencing the phenotype.

The periconceptual, *in utero*, and postnatal developmental environment can have an impact on long-term risk for adult-onset obesity by set point adaptive changes [93]. Early life environmental factors such as breastfeeding have been pointed to protect against childhood obesity [94]. This early life environment was shown to be associated with DNA methylation of *LEP*, a gene that is implicated in appetite regulation and fat metabolism [94]. The human diet suffered profound modifications in the last century, marked by innovations in food technology (processed food), novel ingredients, and artificial vitamins altering the human diet. A "poor" or "excess" nutrition at pre- or periconceptional stages can increase the chance of developing metabolic syndromes (e.g. obesity, hypertension, insulin resistance) in later life, and these developmental origins of adult disease put forward epigenetic inheritance as the possible mechanism in this programming [95, 96]. Dietary factors modulate epigenetic processes and are largely demonstrated to play an important role in cancer and aging [97]. Understanding how these changes may affect health remains still unknown and epigenetic emerge as an important mechanism in this alteration.

Animal studies point that the intrauterine environment, as producing persistent changes in epigenetic marks, influences for a phenotypic change in later life [98]. In animal models, maternal under- and over- nutrition can induce persistent changes in gene expression and metabolism. In humans, the Dutch Hunger Winter in 1944–1945 has been a classic example whether prenatal exposure to famine is associated with differences in methylation levels [99]. Maternal nutrition during gestation can "program" the offspring to develop aspects of metabolic syndrome in later life [20, 100]. Offspring exposed to famine in the last trimester of mother's pregnancy or in the first months of life were less likely to be obese, while those whose mothers had been in early or mid-pregnancy during this famine period were twice as likely to be obese [20, 29]. So, early changes may differ in programming from later changes, fitting with hypotheses of developmental programming and critical epigenetic windows. This observation suggests that the mother's diet can affect descents health, in a way that their children and their grandchildren inherit the same health problems. From a molecular standpoint this occurrence seems to derive from epigenetic mechanisms. If the hypothesis of epigenetic inheritance takes place, it will open new perspectives on the role of the environment in the appearance, disappearance or frequency of certain diseases and conditions.

In the last years, several studies found a possible implication of specific nutrients or other bioactive components with epigenetic processes that regulate gene expression and may contribute to an increased susceptibility to develop obesity [67]. In Wistar rats, the regulation of *LEP* gene expression was shown to be influenced by a high-energy diet by affecting *LEP* gene promoter methylation [101]. Authors suggest that important epiobesigenic genes could be involved in obesity regulating gene expression through epigenetic mechanisms.

DNA methylation requires constant supply of new methyl groups that can be obtained directly from food or chemical precursors such as serine, folate, vitamin B12 (cobalamin), vitamin B6 (pyridoxine), vitamine B2 (riboflavin), methionine, choline, betaine and alcohol, and transferred to DNA and histones through the methyl donor SAM, influencing many genes [52, 102]. Otherwise certain chemical elements stemming from food are required to transport the methyl groups along the body for fixing DNA [103]. For example, a zinc or vitamin B deficiency may lead to alteration in the level of DNA methylation as demonstrated in animal studies [24, 104]. In the case of "Dutch mothers" it is probable that poor food consumption deprived them of nutrients needed for proper methylation pattern in their offspring, and these anomalies were transmitted to the next generation.

Other factors have been shown to influence epigenetic modifications including pollution, drugs, tobacco, stress, and the personal livings [105]. On the other hand, diseases such as allergies, asthma, metabolic syndrome, obesity, diabetes, cancer, etc. may develop epigenetic factors. Other nutrients such as genistein, soy isoflavones, curcumin, sulforaphane, tea polyphenols among others showed to modify the activity of DNA methyltransferases [67].

Histone marks can also be modified through dietary factors by two major ways: altering the abundance and/or efficacy of enzymes responsible for the modification, and altering the availability of the enzyme substrate [52]. Thus, there is a great potential of diet to influence histones. For example, butyrate (*n*-butyric acid), a short-chain naturally occurring fatty acid produced in human by anaerobic bacterial fermentation of carbohydrates (present in dairy products such as Greek feta cheese), and dietary polyphenols have been shown to be histone deacetylase inhibitors [106, 107], whereas histone acetyltransferases enzymes are inhibited by green tea polyphenols and copper [108, 109]. The effect of diets on histone methyltransferases and histone demethylases is less known.

Although the precise mechanism underlying the control of miRNA expression is not well understood, currently, several studies in rodents and humans found that nutrients such as vitamin A, B, D and E, fatty acids, proteins or alcohol alter expression of many miRNAs [110, 111]. Moreover, epigenetic changes could also play a major role in the miRNA expression [112]. Although there are relatively few studies on natural agents that might modulate gene expression by targeting miRNAs, this field is likely to gain significant attention in the future.

Walking to an Epigenetic Food?

Epigenetics has been surely the most rapidly growing field in molecular science in the last 5 years. As we pointed out previously, several nutrients and bioactive food (Table 8.2) could play an active role in the complex interaction between the genome and the epigenome machinery that regulate gene expression [52, 102]. Nutrigenomics is in charge to identify the role of nutrients on gene expression, in which food and nutrition could affect the health [113, 114]. When we eat, we introduce some bioactive dietary components that have molecules carrying information from the external environment [115]. It is not only the quality, but also the quantity of the nutrient ingestion that may affects the process of gene expression [97]. Recently, the term nutriepigenetics appear to define dietary food that modulate gene expression and long term health outcomes through epigenetic modifications [116]. There are several studies that linked early nutrition and poor growth *in utero* with increased risk of several metabolism disorders [98]. Nutrient-gene interaction can modulate epigenetic phenomena by inhibiting enzymes such as DNA methyltransferases and histone deacetylases [52]. It is well know that responses to nutritional interventions such as calorie restriction or low protein diets show a wide range of inter-individual variation [117], which is induced by epigenetic modifications. Chromatin modifications play an important role in the process of DNA repair, DNA recombination and many other chromatin-dependent processes [118]. So, some vitamins or other micro/macro elements can alter these processes affecting chromatin [119].

Dietary interventions based on phytochemicals (e.g. green tea, cruciferous vegetables) are demonstrated to have anticancer epigenetic effects and are also efficacious for preventing or treating the epigenetic changes of other age-associated

Table 8.2 Common examples of nutrients on metabolic effects

	Nutrients	Food sources examples	Animal Model	Metabolic effects	Epigenetic mechanisms	Reference
Methyl donor nutrients	Methionine	Legumes, eggs, fish, garlic, lentils, meat, onions, soy, yogurt, Parmesan, Gruyere	Sheep	Insulin resistance, obesity	DNA methylation	[137]
	Choline	Eggs, fish, legumes, nuts, meat (especially liver)	Rat	Liver steatosis, body weight	DNA methylation, miRNA	[138]
	Betaine	Spinach, cereals, beets, quinoa, fish	Rat	Liver steatosis, body weight	DNA methylation, miRNA expression	[138]
Vitamins	Vitamin B12	Fish, shellfish, meat (especially liver), eggs, milk, poultry	Sheep	Insulin resistance, obesity	DNA methylation	[137]
			Rat	Liver steatosis, body weight	DNA methylation, miRNA expression	[138]
	Folate (B9)	Dried apricots, nuts, asparagus, broccoli, yeast, soy, spinach	Sheep	Insulin resistance, adiposity	DNA methylation	[137]
	Vitamin B6	Poultry, liver (beef, lamb, veal), bananas, various forms of cabbage, tomatoes, spinach, potatoes	Rat	Obesity, body weight, insulin resistance	DNA methylation	[139]
	Vitamin A	Cod liver oil, liver (beef, turkey, chicken), fish, spinach, carrot, broccoli, sweet potato	Rat	Heart defect	DNA methylation	[140]
Fatty acids	Saturated fat	Cheese, pizza, burgers	Rat	Body weight, Energy metabolism	DNA methylation	[141]
	Lipids	Lard, cheese, palm oil, fried food	Rat	Adiposity, Obesity	DNA methylation	[101]
	Butyrate acetate propionate	Lard, soybean oil.	Human	Gut microbiota	DNA methylation	[142]

Table 8.2 (continued)

	Nutrients	Food sources examples	Animal Model	Metabolic effects	Epigenetic mechanisms	Reference
Micronutrients	Zinc	Seafood, beef, lamb, wheat germ, spinach, nuts, beans	Rat	Body weight	Fetal programming	[143]
	Chromium	Broccoli, barley, green beans, tomatoes	Rat	adiposity	Fetal programming	[144]
	Magnesium	Cereals, rice, bread, spinach, fish, seafood	Rat	Adiposity	Fetal programming	[145]
Bioactive food component	Polyphenols	Apples	Rat	Adiposity	DNA methylation	[146]
	Genistein	Soybean, soy products	Mice	Body weight	DNA methylation	[147]
	Epigallocatechin gallate	White and green tea	Human	Obesity	Histone deacetylase	[120]
	Resveratrol	Wine, red grapes, berries, peanuts	Rat	Obesity, fat metabolism	miRNA expression	[148]

diseases, with an efficacy for prevention and treatment of associated epigenetic aberrations that are readily reversible [97]. Many bioactive components with epigenetic properties can be found. For example, epigallocatechin gallate (EGCG), a potent anti-inflammatory agent and the most abundant component of green tea, has been suggested to have a role as a preventive agent in obesity, cancer and other disorders inducing histone modifications [120]. Sulforaphane, an aliphatic isothiocyanate present in cruciferous vegetables such as broccoli, cauliflower and cabbage, is known for its association with adipocytes and has been shown that can inhibit Dnmt1, 3A and histone deactylases activity [97]. Dietary intervention on sulforaphane suggest an anti-obesity activity by inhibiting adipogenesis and suppressing lipogenesis [121]. An anti-obesity effect with reduction of body weight and abdominal obesity was also found in dietary intervention with resveratrol, present in red grapes [122], playing a role in epigenetic changes [123]. Soybean products are a good source of proteins that contains isoflavones, including genistein; data obtained in non-human primates showed epigenetic modifications associated with dietary interventions with soy protein that may potentially alter the risk of obesity [124]. There are some evidences demonstrating that diets rich in polyphenols contained in fruits and vegetables promote health and attenuate, or delay, the onset of cardiovascular diseases through epigenetic changes [115]. Other study has suggested the use of soybean in prevention to repress early breast tumorigenesis [125].

The DNA methylation profile can change under the influence of genetic variation [126]. Methylenetetrahydrofolate reductase (*MTHFR*) is a key enzyme in the one-carbon metabolism, which serves as methyl donor for methionine, precursor of SAM [127, 128]. The common missense mutation C677T in *MTHFR* gene has been demonstrated in gene-nutrient interactions, and individuals with the mutant T allele show less enzyme activity compared with the wild-type homozygous state (CC) [129]. In 2002, Friso et al. [128] demonstrated the influences of the *MTHFR* C677T polymorphism in DNA methylation status through an interaction with folate status.

This data suggests that, in the future, nutritional agents or "epigenetic foods" containing bioactive compounds could have the capacity to modulate DNA methylation, histone modifications, or miRNA expression. However, the actual molecular interactions of nutrients with biological systems remain mostly speculative.

Epigenetics: The Return to Lamarckism?

The possibility that some epigenetic marks could be acquired and transmitted from parents to children has a deliciously Lamarckian flavour. This idea is irresistible as a potential "cure" to genetic determinism. In 1809, the book "*Philosophie Zoologique*" by Jean-Baptiste Lamarck was published explaining his theory of the progressive transformation of species as an alternative to creationism. For Lamarck, the evolution of species is a proven fact. His theory was based on the influence of environment in the development of organs, the modification of the organs by the use or non-use and the transmission of these changes from one generation to another. Sure, the examples chosen to illustrate his theory were not the most appropriate: “the giraffe becomes obliged to eat the leaves of the tallest trees, and so her neck elongates” contributed to ridicule his work. Nevertheless, the idea of Lamarck could not be totally wrong. Recently, and as stated above, several studies demonstrated environment as a cause of phenotypic changes, by only modifying the methylation pattern (for the example about honey bees), without modifying the DNA sequence. These studies showed how environmental factors could modify the Mendelian inheritance as a result of these epigenetic factors.

To Lamarckism the transformation of species occurs through the “inheritance of acquired characters”, and by these modifications all organisms would become adapted to their environments as those environments changed. According to Darwinism, the variations (i. e., mutations) are random and are selected according to their advantage for the carrier. Lamarck was refuted because the experience held by August Weismann which consisted of the mutilation of the tail of rats [130]. Over several generations, the descendants had not “acquired” the absence of a tail. For Weismann, individual adaptation could not be inherited because of the existence of an insuperable barrier. Certain experiences came to shake the classical view of hereditary mechanisms. Indeed, there is a hereditary “no gene” that is not registered in the DNA chain. Epigenetic mechanisms regulate gene expression and function as a gene memory that could be transmitted from parents to children. The changes in our diet in the last century could have an important influence in some epigenetic mechanisms increasing the susceptibility to develop an obese phenotype. So, we need to reinvent the Lamarckism, and thus speak in epigenetics as neo-Lamarckism?

When Epigenetics Open Us New Perspectives

Epigenetics shows that we are not simply determined by our genes. On the contrary, we are able by our behaviors to “change them”. Inappropriate behavior will lead to epigenetic changes responsible for certain diseases such as diabetes, obesity, cardiovascular diseases, cancers, asthma and other. Instead, behaviors for healthy lifestyle will produce “good” changes in the expression of our genes, allowing us to stay in the path of healing and good health. One of the most important environmental factors to which we are exposed along our lifetime is our diet. Effectively, the ingestion of nutrients and bioactive food components can influence and modulate epigenetic phenomena affecting gene expression. It is well known that some nutrients or other lifestyle factors affect physiologic and pathologic processes in our body. Certain nutrients can modulate epigenetic changes in the epigenetic genome by inhibiting enzymes such as DNA methyltransferases and histone deacetylases. As we discussed previously, epigenetics patterns such as DNA methylation or histone modifications are very important in gene expression. DNA methylation is important to control gene expression and chromosome structure, whereas histone modifications modulate chromatin structure controlling DNA transcription, replication, recombination, and repair. Until now, our knowledge about the roles of specific epigenetic marks, proteins and their interactions remain unclear. As proposed by Zaidi et al. [131] it can exist an "architectural epigenetics", describing all combination of mitotically inheritable epigenetic mechanisms which control gene expression. However, there are no yet convincing proofs between the interactions of epigenetic mechanisms as key regulators of gene transcription in responses to altered environmental signals.

Epigenetics can open us a new door inside the treatment to obesity, as the diverse nutritional exposures could have the potential to influence epigenetic regulation in some genomic regions, potentially in a tissue- and cell-type specific manner. So, identification of individuals with a specific methylation profile, susceptible to weight gain could be used to prevent and monitor their evolution with a specific diet. The understanding of how nutrients affect epigenetic mechanisms and how deregulation contributes to obesity are important to create a specific and individualized treatment. We are only in the beginning of the understanding of the contribution of epigenetic marks to the increased risk of developing an obese phenotype. Thus, it is a new exciting challenge to unveil epigenetic mechanisms associated with the development of human obesity and to elaborate specific therapeutic concepts into new directions.

References

1. Swinburn B, Sacks G, Hall KD et al (2011) The global obesity pandemic: shaped by global drivers and local environments. Lancet 378:804–814
2. Herrera BM, Keildson S, Lindgren CM (2011) Genetics and epigenetics of obesity. Maturitas 69:41–49
3. Stunkard AJ, Foch TT, Hrubec Z (1986) A twin study of human obesity. JAMA 256:51–54
4. Stunkard AJ, Sørensen TI, Hanis C et al (1986) An adoption study of human obesity. N Engl J Med 314:193–198
5. Hjelmborg JB, Fagnani C, Silventoinen K et al (2008) Genetic influences on growth traits of BMI: a longitudinal study of adult twins. Obesity (Silver Spring) 16:847–852
6. Wardle J, Carnell S, Haworth CM, Plomin R (2008) Evidence for a strong genetic influence on childhood adiposity despite the force of the obesogenic environment. Am J Clin Nutr 87:398–404
7. Silventoinen K, Rokholm B, Kaprio J, Sørensen TIA (2010) The genetic and environmental influences on childhood obesity: a systematic review of twin and adoption studies. Int J Obes (Lond) 34:29–40
8. Berndt SI, Gustafsson S, Mägi R et al (2013) Genome-wide meta-analysis identifies 11 new loci for anthropometric traits and provides insights into genetic architecture. Nat Genet 45:501–512
9. Frayling TM, Timpson NJ, Weedon MN et al (2007) A common variant in the FTO gene is associated with body mass index and predisposes to childhood and adult obesity. Science 316:889–894
10. Speliotes EK, Willer CJ, Berndt SI et al (2010) Association analyses of 249,796 individuals reveal 18 new loci associated with body mass index. Nat Genet 42:937–948
11. Loke KY, Lin JB, Mabel DY (2008) 3rd College of paediatrics and child health lecture-the past, the present and the shape of things to come. Ann Acad Med Singap 37:429–434
12. Waddington CH (2014) The strategy of the genes: a discussion of some aspects of theoretical biology. Routledge library editions: 20th Century Science

13. Morange M (2002) The relations between genetics and epigenetics: a historical point of view. Ann N Y Acad Sci 981:50–60
14. Campión J, Milagro FI, Martínez JA (2009) Individuality and epigenetics in obesity. Obes Rev 10:383–392
15. Ling C, Groop L (2009) Epigenetics: a molecular link between environmental factors and type 2 diabetes. Diabetes 58:2718–2725
16. Girardot M, Cavaillé J, Feil R (2012) Small regulatory RNAs controlled by genomic imprinting and their contribution to human disease. Epigenetics 7:1341–1348
17. Holliday R (1990) DNA methylation and epigenetic inheritance. Philos Trans R Soc Lond B Biol Sci 326:329–338
18. Riggs AD, Martienssen RA, Russo VEA (1996) Introduction. In: Riggs AD, Martienssen RA, Russo VEA (eds) Epigenetic mech, Gene regul cold spring harbor laboratory Press: Plainview, New York, p. 692
19. Feil R, Fraga MF (2011) Epigenetics and the environment: emerging patterns and implications. Nat Rev Genet 13:97–109
20. Heijmans BT, Tobi EW, Stein AD et al (2008) Persistent epigenetic differences associated with prenatal exposure to famine in humans. Proc Natl Acad Sci U S A 105:17046–17049
21. Kucharski R, Maleszka J, Foret S, Maleszka R (2008) Nutritional control of reproductive status in honeybees via DNA methylation. Science 319:1827–1830
22. Lyko F, Foret S, Kucharski R et al (2010) The honey bee epigenomes: differential methylation of brain DNA in queens and workers. PLoS Biol 8:e1000506
23. Kim D-H, Doyle MR, Sung S, Amasino RM (2009) Vernalization: winter and the timing of flowering in plants. Annu Rev Cell Dev Biol 25:277–299
24. Wolff GL, Kodell RL, Moore SR, Cooney CA (1998) Maternal epigenetics and methyl supplements affect agouti gene expression in Avy/a mice. FASEB J 12:949–957
25. Bird A (2007) Perceptions of epigenetics. Nature 447:396–398
26. Rakyan VK, Blewitt ME, Druker R et al (2002) Metastable epialleles in mammals. Trends Genet 18:348–351
27. Waterland RA, Jirtle RL (2003) Transposable elements: targets for early nutritional effects on epigenetic gene regulation. Mol Cell Biol 23:5293–5300
28. Khosla S, Dean W, Brown D et al (2001) Culture of preimplantation mouse embryos affects fetal development and the expression of imprinted genes. Biol Reprod 64:918–926
29. Tobi EW, Lumey LH, Talens RP et al (2009) DNA methylation differences after exposure to prenatal famine are common and timing- and sex-specific. Hum Mol Genet 18:4046–4053
30. Hirasawa R, Feil R (2010) Genomic imprinting and human disease. Essays Biochem 48:187–200
31. Saetrom P, Snøve O, Rossi JJ (2007) Epigenetics and microRNAs. Pediatr Res 61:17R–23R
32. Kim JK, Samaranayake M, Pradhan S (2009) Epigenetic mechanisms in mammals. Cell Mol Life Sci 66:596–612
33. Handel AE, Ebers GC, Ramagopalan SV (2010) Epigenetics: molecular mechanisms and implications for disease. Trends Mol Med 16:7–16
34. Relton CL, Davey Smith G (2010) Epigenetic epidemiology of common complex disease: prospects for prediction, prevention, and treatment. PLoS Med 7:e1000356
35. Bird A (2002) DNA methylation patterns and epigenetic memory. Genes Dev 16:6–21
36. Campión J, Milagro FI, Martínez J a (2009) Individuality and epigenetics in obesity. Obes Rev 10:383–392
37. Law JA, Jacobsen SE (2010) Establishing, maintaining and modifying DNA methylation patterns in plants and animals. Nat Rev Genet 11:204–220
38. Costello JF, Plass C (2001) Methylation matters. J Med Genet 38:285–303
39. Chen T, Li E (2006) Establishment and maintenance of DNA methylation patterns in mammals. Curr Top Microbiol Immunol 301:179–201
40. Lillycrop K a, Burdge GC (2011) Epigenetic changes in early life and future risk of obesity. Int J Obes (Lond) 35:72–83

41. Reik W, Lewis A (2005) Co-evolution of X-chromosome inactivation and imprinting in mammals. Nat Rev Genet 6:403–410
42. Zhang Y, Reinberg D (2001) Transcription regulation by histone methylation: interplay between different covalent modifications of the core histone tails. Genes Dev 15:2343–2360
43. Van Holde KE (1989) Chromatin. New York: Springer-Verlag. p. 497
44. Olins AL, Olins DE (1974) Spheroid chromatin units (v bodies). Science 183:330–332
45. Khorasanizadeh S (2004) The nucleosome: from genomic organization to genomic regulation. Cell 116:259–272
46. Zhou VW, Goren A, Bernstein BE (2011) Charting histone modifications and the functional organization of mammalian genomes. Nat Rev Genet 12:7–18
47. Zhu Q, Wani AA (2010) Histone modifications: crucial elements for damage response and chromatin restoration. J Cell Physiol 223:283–288
48. Jenuwein T, Allis CD (2001) Translating the histone code. Science 293:1074–1080
49. Shiio Y, Eisenman RN (2003) Histone sumoylation is associated with transcriptional repression. Proc Natl Acad Sci U S A 100:13225–13230
50. Egger G, Liang G, Aparicio A, Jones PA (2004) Epigenetics in human disease and prospects for epigenetic therapy. Nature 429:457–463
51. Youngson NA, Morris MJ (2013) What obesity research tells us about epigenetic mechanisms. Philos Trans R Soc Lond B Biol Sci 368:20110337
52. McKay JA, Mathers JC (2011) Diet induced epigenetic changes and their implications for health. Acta Physiol (Oxf) 202:103–118
53. Bladé C, Baselga-Escudero L, Salvadó MJ, Arola-Arnal A (2013) miRNAs, polyphenols, and chronic disease. Mol Nutr Food Res 57:58–70
54. Sevignani C, Calin GA, Siracusa LD, Croce CM (2006) Mammalian microRNAs: a small world for fine-tuning gene expression. Mamm Genome 17:189–202
55. Pillai RS, Bhattacharyya SN, Filipowicz W (2007) Repression of protein synthesis by miRNAs: how many mechanisms? Trends Cell Biol 17:118–126
56. Bentwich I, Avniel A, Karov Y et al (2005) Identification of hundreds of conserved and non-conserved human microRNAs. Nat Genet 37:766–770
57. Martinelli R, Nardelli C, Pilone V et al (2010) miR-519d overexpression is associated with human obesity. Obesity (Silver Spring) 18:2170–2176
58. Williams MD, Mitchell GM (2012) MicroRNAs in insulin resistance and obesity. Exp Diabetes Res 2012:484696.
59. Teperino R, Lempradl A, Pospisilik JA (2013) Bridging epigenomics and complex disease: the basics. Cell Mol Life Sci 70:1609–1621
60. McGregor RA, Choi MS (2011) microRNAs in the regulation of adipogenesis and obesity. Curr Mol Med 11:304–316
61. Dehwah MAS, Xu A, Huang Q (2012) MicroRNAs and type 2 diabetes/obesity. J Genet Genomics 39:11–18
62. Wang J, Hevi S, Kurash JK et al (2009) The lysine demethylase LSD1 (KDM1) is required for maintenance of global DNA methylation. Nat Genet 41:125–129. doi:10.1038/ng.268
63. Zhao Q, Rank G, Tan YT et al (2009) PRMT5-mediated methylation of histone H4R3 recruits DNMT3A, coupling histone and DNA methylation in gene silencing. Nat Struct Mol Biol 16:304–311
64. Byun H-M, Siegmund KD, Pan F et al (2009) Epigenetic profiling of somatic tissues from human autopsy specimens identifies tissue- and individual-specific DNA methylation patterns. Hum Mol Genet 18:4808–4817
65. Lister R, Pelizzola M, Dowen RH et al (2009) Human DNA methylomes at base resolution show widespread epigenomic differences. Nature 462:315–322
66. Drong a W, Lindgren CM, McCarthy MI (2012) The genetic and epigenetic basis of type 2 diabetes and obesity. Clin Pharmacol Ther 92:707–715
67. Milagro FI, Mansego ML, De Miguel C, Martínez JA Dietary factors, epigenetic modifications and obesity outcomes: progresses and perspectives. Mol Aspects Med 34:782–812

68. Bell CG, Finer S, Lindgren CM et al (2010) Integrated genetic and epigenetic analysis identifies haplotype-specific methylation in the FTO type 2 diabetes and obesity susceptibility locus. PLoS One 5:e14040
69. Almén MS, Jacobsson J, Moschonis G et al (2012) Genome wide analysis reveals association of a FTO gene variant with epigenetic changes. Genomics 99:132–137
70. Toperoff G, Aran D, Kark JD et al (2012) Genome-wide survey reveals predisposing diabetes type 2-related DNA methylation variations in human peripheral blood. Hum Mol Genet 21:371–383
71. Gerken T, Girard CA, Tung Y-CL et al (2007) The obesity-associated FTO gene encodes a 2-oxoglutarate-dependent nucleic acid demethylase. Science 318:1469–1472
72. Xu X, Su S, Barnes VA et al (2013) A genome-wide methylation study on obesity: differential variability and differential methylation. Epigenetics 8:522–533
73. Wang X, Zhu H, Snieder H et al (2010) Obesity related methylation changes in DNA of peripheral blood leukocytes. BMC Med 8:87
74. Tobi EW, Heijmans BT, Kremer D et al (2011) DNA methylation of IGF2, GNASAS, INSIGF and LEP and being born small for gestational age. Epigenetics 6:171–176
75. Brøns C, Jacobsen S, Nilsson E et al (2010) Deoxyribonucleic acid methylation and gene expression of PPARGC1A in human muscle is influenced by high-fat overfeeding in a birth-weight-dependent manner. J Clin Endocrinol Metab 95:3048–3056
76. Crujeiras AB, Campion J, Díaz-Lagares A et al (2013) Association of weight regain with specific methylation levels in the NPY and POMC promoters in leukocytes of obese men: a translational study. Regul Pept 186:1–6
77. Lan MY, Chang YY, Chen WH et al (2009) Serotonin transporter gene promoter polymorphism is associated with body mass index and obesity in non-elderly stroke patients. J Endocrinol Invest 32:119–122
78. Calati R, De Ronchi D, Bellini M, Serretti A (2011) The 5-HTTLPR polymorphism and eating disorders: a meta-analysis. Int J Eat Disord 44:191–199
79. Zhao J, Goldberg J, Vaccarino V (2013) Promoter methylation of serotonin transporter gene is associated with obesity measures: a monozygotic twin study. Int J Obes (Lond) 37:140–145
80. Ollikainen M, Smith KR, Joo EJ-H et al (2010) DNA methylation analysis of multiple tissues from newborn twins reveals both genetic and intrauterine components to variation in the human neonatal epigenome. Hum Mol Genet 19:4176–4188
81. Relton CL, Groom A, St Pourcain B et al (2012) DNA methylation patterns in cord blood DNA and body size in childhood. PLoS One 7:e31821
82. Godfrey KM, Sheppard A, Gluckman PD et al (2011) Epigenetic gene promoter methylation at birth is associated with child's later adiposity. Diabetes 60:1528–1534
83. Soubry A, Schildkraut JM, Murtha A et al (2013) Paternal obesity is associated with IGF2 hypomethylation in newborns: results from a Newborn Epigenetics Study (NEST) cohort. BMC Med 11:29
84. Soubry A, Murphy SK, Wang F et al (2013) Newborns of obese parents have altered DNA methylation patterns at imprinted genes. Int J Obes (Lond) doi:10.1038/ijo.2013.193
85. Souren NYP, Tierling S, Fryns J-P et al (2011) DNA methylation variability at growth-related imprints does not contribute to overweight in monozygotic twins discordant for BMI. Obesity (Silver Spring) 19:1519–1522
86. Drong AW, Nicholson G, Hedman AK et al (2013) The presence of methylation quantitative trait loci indicates a direct genetic influence on the level of DNA methylation in adipose tissue. PLoS ONE 8:e55923
87. Rönn T, Volkov P, Davegårdh C et al (2013) A six months exercise intervention influences the genome-wide DNA methylation pattern in human adipose tissue. PLoS Genet 9:e1003572
88. Meerson A, Traurig M, Ossowski V et al (2013) Human adipose microRNA-221 is upregulated in obesity and affects fat metabolism downstream of leptin and TNF-. Diabetologia 56:1971–1979

89. Xie H, Lim B, Lodish HF (2009) MicroRNAs induced during adipogenesis that accelerate fat cell development are downregulated in obesity. Diabetes 58:1050–1057
90. Takanabe R, Ono K, Abe Y et al (2008) Up-regulated expression of microRNA-143 in association with obesity in adipose tissue of mice fed high-fat diet. Biochem Biophys Res Commun 376:728–732
91. Klöting N, Berthold S, Kovacs P et al (2009) MicroRNA expression in human omental and subcutaneous adipose tissue. PLoS One 4:e4699
92. Ortega FJ, Moreno-Navarrete JM, Pardo G et al (2010) MiRNA expression profile of human subcutaneous adipose and during adipocyte differentiation. PLoS ONE 5:e9022
93. Godfrey KM, Gluckman PD, Hanson MA (2010) Developmental origins of metabolic disease: life course and intergenerational perspectives. Trends Endocrinol Metab 21:199–205
94. Obermann-Borst S, Eilers PHC, Tobi EW et al (2013) Duration of breastfeeding and gender are associated with methylation of the LEPTIN gene in very young children. Pediatr Res 74:344–349
95. Dyer JS, Rosenfeld CR (2011) Metabolic imprinting by prenatal, perinatal, and postnatal overnutrition: a review. Semin Reprod Med 29:266–276
96. Faulk C, Dolinoy DC (2011) Timing is everything: the when and how of environmentally induced changes in the epigenome of animals. Epigenetics 6:791–797
97. Tollefsbol TO (2014) Dietary epigenetics in cancer and aging. Cancer Treat Res 159:257–267
98. Seki Y, Williams L, Vuguin PM, Charron MJ (2012) Minireview: epigenetic programming of diabetes and obesity: animal models. Endocrinology 153:1031–1038
99. Lumey LH, Stein AD, Kahn HS et al (2007) Cohort profile: the Dutch hunger winter families study. Int J Epidemiol 36:1196–1204
100. Ravelli GP, Stein ZA, Susser MW (1976) Obesity in young men after famine exposure in utero and early infancy. N Engl J Med 295:349–353
101. Milagro FI, Campión J, García-Díaz DF et al (2009) High fat diet-induced obesity modifies the methylation pattern of leptin promoter in rats. J Physiol Biochem 65:1–9
102. Ross SA (2003) Diet and DNA methylation interactions in cancer prevention. Ann N Y Acad Sci 983:197–207
103. Szarc velSK, Ndlovu MN, Haegeman G, Vanden Berghe W (2010) Nature or nurture: let food be your epigenetic medicine in chronic inflammatory disorders. Biochem Pharmacol 80:1816–1832
104. Wallwork JC, Duerre JA (1985) Effect of zinc deficiency on methionine metabolism, methylation reactions and protein synthesis in isolated perfused rat liver. J Nutr 115:252–262
105. Franks PW, Ling C (2010) Epigenetics and obesity: the devil is in the details. BMC Med 8:88
106. Link A, Balaguer F, Goel A (2010) Cancer chemoprevention by dietary polyphenols: promising role for epigenetics. Biochem Pharmacol 80:1771–1792
107. Steliou K, Boosalis MS, Perrine SP et al (2012) Butyrate histone deacetylase inhibitors. Biores Open Access 1:192–198
108. Lin C, Kang J, Zheng R (2005) Oxidative stress is involved in inhibition of copper on histone acetylation in cells. Chem Biol Interact 151:167–176
109. Choi K-C, Jung MG, Lee Y-H et al (2009) Epigallocatechin-3-gallate, a histone acetyltransferase inhibitor, inhibits EBV-induced B lymphocyte transformation via suppression of RelA acetylation. Cancer Res 69:583–592
110. Mathers JC, Strathdee G, Relton CL (2010) Induction of epigenetic alterations by dietary and other environmental factors. Adv Genet 71:3–39
111. Parasramka MA, Ho E, Williams DE, Dashwood RH (2012) MicroRNAs, diet, and cancer: new mechanistic insights on the epigenetic actions of phytochemicals. Mol Carcinog 51:213–230
112. Kong A-NT, Zhang C, Su Z-Y (2013) Targeting epigenetics for cancer prevention by dietary cancer preventive compounds-the case of miRNA. Cancer Prev Res (Phila) 6:622–624

113. Phillips CM (2013) Nutrigenetics and metabolic disease: current status and implications for personalised nutrition. Nutrients 5:32–57
114. Neeha VS, Kinth P (2013) Nutrigenomics research: a review. J Food Sci Technol 50:415–428
115. Barnes S (2008) Nutritional genomics, polyphenols, diets, and their impact on dietetics. J Am Diet Assoc 108:1888–1895
116. Amandio V (2012) Nutriepigenetics and related topics: from cell mechanisms to possible therapeutic relevance. Cell Biol: Res Ther 1:2
117. Abete I, Navas-Carretero S, Marti A, Martinez JA (2012) Nutrigenetics and nutrigenomics of caloric restriction. Prog Mol Biol Transl Sci 108:323–346
118. Kouzarides T (2007) Chromatin modifications and their function. Cell 128:693–705
119. Ferguson LR, Karunasinghe N, Philpott M (2004) Epigenetic events and protection from colon cancer in New Zealand. Environ Mol Mutagen 44:36–43
120. Yun J-M, Jialal I, Devaraj S (2010) Effects of epigallocatechin gallate on regulatory T cell number and function in obese v. lean volunteers. Br J Nutr 103:1771–1777
121. Choi K-M, Lee Y-S, Kim W et al (2014) Sulforaphane attenuates obesity by inhibiting adipogenesis and activating the AMPK pathway in obese mice. J Nutr Biochem 25:201–207
122. Baek S-H, Chung H-J, Lee H-K et al (2014) Treatment of obesity with the resveratrol-enriched rice DJ-526. Sci Rep 4:3879
123. Martin SL, Hardy TM, Tollefsbol TO (2013) Medicinal chemistry of the epigenetic diet and caloric restriction. Curr Med Chem 20:4050–4059
124. Howard TD, Ho S-M, Zhang L et al (2011) Epigenetic changes with dietary soy in cynomolgus monkeys. PLoS One 6:e26791
125. Li Y, Chen H, Hardy TM, Tollefsbol TO (2013) Epigenetic regulation of multiple tumor-related genes leads to suppression of breast tumorigenesis by dietary genistein. PLoS ONE 8:e54369
126. Eichten SR, Briskine R, Song J et al (2013) Epigenetic and genetic influences on DNA methylation variation in maize populations. Plant Cell 25:2783–2797
127. Friso S, Choi S-W (2002) Gene-nutrient interactions and DNA methylation. J Nutr 132:2382S–2387S
128. Friso S, Choi S-W, Girelli D et al (2002) A common mutation in the 5,10-methylenetetrahydrofolate reductase gene affects genomic DNA methylation through an interaction with folate status. Proc Natl Acad Sci U S A 99:5606–5611
129. Stern LL, Mason JB, Selhub J, Choi SW (2000) Genomic DNA hypomethylation, a characteristic of most cancers, is present in peripheral leukocytes of individuals who are homozygous for the C677T polymorphism in the methylenetetrahydrofolate reductase gene. Cancer Epidemiol Biomarkers Prev 9:849–853
130. Weismann A (2006) Essays upon heredity and kindred biological problems, vol 2. p 240
131. Zaidi SK, Young DW, Montecino M et al (2010) Architectural epigenetics: mitotic retention of mammalian transcriptional regulatory information. Mol Cell Biol 30:4758–4766
132. Stepanow S, Reichwald K, Huse K et al (2011) Allele-specific, age-dependent and BMI-associated DNA methylation of human MCHR1. PLoS ONE 6(5):e17711
133. Milagro FI, Gómez-Abellán P, Campión J, Martínez JA, Ordovás JM, Garaulet M (2012) CLOCK, PER2 and BMAL1 DNA methylation: association with obesity and metabolic syndrome characteristics and monounsaturated fat intake. Chronobiol Int 29(9):1180–1194
134. Liu X, Chen Q, Tsai HJ et al (2014) Maternal preconception body mass index and offspring cord blood DNA methylation: exploration of early life origins of disease. Environ Mol Mutagen 55(3):223–230
135. Nardelli C, Iaffaldano L, Ferrigno M et al (2014) Characterization and predicted role of the microRNA expression profile in amnion from obese pregnant women. Int J Obes (Lond) 38(3):466–469
136. Moleres A, Campión J, Milagro FI et al (2013) Differential DNA methylation patterns between high and low responders to a weight loss intervention in overweight or obese adolescents: the EVASYON study. FASEB J 27(6):2504–2512

137. Sinclair KD, Allegrucci C, Singh R et al (2007) DNA methylation, insulin resistance, and blood pressure in offspring determined by maternal periconceptional B vitamin and methionine status. Proc Natl Acad Sci U S A 104(49):19351–19356
138. Cordero P, Campion J, Milagro FI, Martinez J (2013) Transcriptomic and epigenetic changes in early liver steatosis associated to obesity: effect of dietary methyl donor supplementation. Mol Genet Metab 110(3):388–395
139. Cho CE (2014) Role of methyl group vitamins in hypothalamic development of food intake regulation in Wistar rats. Appl Physiol Nutr Metab 39(7):844
140. Feng Y, Zhao LZ, Hong L, Shan C, Shi W, Cai W (2013) Alteration in methylation pattern of GATA-4 promoter region in vitamin A-deficient offspring's heart. J Nutr Biochem 24(7):1373–1380
141. Lomba A, Martínez JA, García-Díaz DF, Paternain L, Marti A, Campión J, Milagro FI (2010) Weight gain induced by an isocaloric pair-fed high fat diet: a nutriepigenetic study on FASN and NDUFB6 gene promoters. Mol Genet Metab 101(2–3):273–278
142. Remely M, Aumueller E, Merold C, Dworzak S, Hippe B, Zanner J, Pointner A (2014) Effects of short chain fatty acid producing bacteria on epigenetic regulation of FFAR3 in type 2 diabetes and obesity. Gene 537(1):85–92
143. Jou MY, Philipps AF, Lönnerdal B (2010) Maternal zinc deficiency in rats affects growth and glucose metabolism in the offspring by inducing insulin resistance postnatally. J Nutr 140(9):1621–1627
144. Padmavathi IJ, Rao KR, Venu L, Ganeshan M, Kumar KA, Rao ChN, Harishankar N, Ismail A, Raghunath M (2010) Chronic maternal dietary chromium restriction modulates visceral adiposity: probable underlying mechanisms. Diabetes 59(1):98–104
145. Venu L, Padmavathi IJ, Kishore YD, Bhanu NV, Rao KR, Sainath PB, Ganeshan M, Raghunath M (2008) Long-term effects of maternal magnesium restriction on adiposity and insulin resistance in rat pups. Obesity (Silver Spring) 16(6):1270–1276
146. Boqué N, de la Iglesia R, de la Garza AL et al (2013) Prevention of diet-induced obesity by apple polyphenols in Wistar rats through regulation of adipocyte gene expression and DNA methylation patterns. Mol Nutr Food Res 57(8):1473–1478
147. Dolinoy DC1, Weidman JR, Waterland RA, Jirtle RL (2006) Maternal genistein alters coat color and protects Avy mouse offspring from obesity by modifying the fetal epigenome. Environ Health Perspect 114(4):567–572
148. Nagao K, Jinnouchi T, Kai S, Yanagita T (2013) Effect of dietary resveratrol on the metabolic profile of nutrients in obese OLETF rats. Lipids Health Dis 12:8

Chapter 9
MicroRNAs in Obesity and Metabolism

Lígia Sousa-Ferreira, Luís Pereira de Almeida and Cláudia Cavadas

MicroRNAs are small, endogenous, non-coding RNAs that play an important regulatory role in cell physiology because of their inhibitory effect in gene expression. MicroRNAs regulate several metabolic processes including adipocyte differentiation, insulin production and secretion, glucose metabolism and insulin resistance. In this context, deregulation of the miRNA pathway, including changes in the levels of miRNAs and target mRNAs, results in defective or distorted cellular mechanisms that contribute to the pathogenesis of obesity and related complications.

A growing body of evidence shows that miRNA are important regulators of several metabolic processes including adipocyte differentiation, insulin production and secretion, glucose metabolism and insulin resistance. More importantly, it has been reported that deregulation of the miRNA pathway is associated to obesity and related conditions such as diabetes. Accordingly, global miRNAs profiling studies in rodents and humans show that obesity, high-fat diet and diabetes lead to alterations in the expression pattern of specific miRNAs in different metabolic tissues namely, adipose tissue, muscle, liver and pancreas. Moreover, changes in miRNAs and target mRNAs levels result in defective or distorted cellular mechanisms that may contribute to the pathogenesis of obesity and related complications. In this context, miRNA modulation emerges as a possible therapeutic option to reestablish the regulatory balance in biological pathways affect by obesity. Another interesting concept is the utilization of plasma miRNA levels as biomarkers for obesity and diabetes.

This chapter will focus on the role of miRNAs and their targets in the cellular mechanisms regulating metabolism and the involvement of miRNAs in the pathogenesis of obesity and related diseases.

C. Cavadas (✉) · L. Pereira de Almeida
Center For Neuroscience and Cell Biology & Faculty of Pharmacy,
University of Coimbra, Coimbra, Portugal
e-mail: ccavadas@gmail.com

L. Sousa-Ferreira
CNC-Center for Neuroscience and Cell Biology, University of Coimbra, Coimbra, Portugal

C. Nóbrega, R. Rodriguez-López (eds.), *Molecular Mechanisms Underpinning the Development of Obesity,* DOI 10.1007/978-3-319-12766-8_9

MicroRNAs

MicroRNAs (miRNAs; miRs) are small, endogenous, non-coding RNAs that play an important regulatory role in cell physiology because of their inhibitory effect in gene expression [1–3]. Since they were first described [4, 5], hundreds of individual miRNAs have been identified in animal and plant cells, and associated to a broad range of biological functions, such as tissue development, cell division, cell fate decision, apoptosis, inflammation and immune response [1]. Additionally, a growing body of evidence shows that miRNAs are crucial regulators of cellular processes controlling metabolism and homeostasis, such as adipocyte differentiation, insulin production and secretion, glucose metabolism and hepatic lipid metabolism [6]. Given the critical role of miRNAs in these mechanisms, it is understandable that a deregulation of the miRNA pathway participates in the pathogenesis of diseases like obesity and diabetes.

The 21–25 nucleotide miRNAs operate by binding to the 3' untranslated region (3' UTR) of target mRNA based on nucleotide sequence, and they achieve gene silencing by promoting the cleavage or translational repression of mRNA transcripts [7]. The regulation of biological mechanisms by miRNAs is a complex system because a single mRNA can be simultaneously repressed by more than one miRNA specie and an individual miRNA can target hundreds of mRNA transcripts [8, 9]. Moreover, the degree of repression is correlated with the number of miRNA binding sites and the amount of target mRNA and miRNAs available within the cell [8]. Importantly, miRNA expression levels are tissue and developmental stage specific [1].

The miRNA biosynthesis pathway (Fig. 9.1) begins with the transcription of gene-encoded miRNAs to primary miRNAs (pri-miRNA), which are long polyadenylated mRNAs containing a hairpin-loop (duplex structure) [10, 11]. Then, pri-miRNAs are processed by Drosha (RNAse enzyme) to a 60–70 nucleotide pre-miRNA that includes the loop structure [12, 13]. The pre-miRNA is exported to the cytoplasm and further processed by endonuclease Dicer [14, 15], resulting in the small double-stranded miRNA. The mature miRNA is constituted by the antisense strand (complementary to the target mRNA sequence), which is loaded to the miR-Induced Silencing Complex (miRISC) [16]. The miRISC complex consists of a number of enzymes including proteins of the Argonaute family, of which AGO2 is responsible for cleaving the mRNA transcripts [16, 17].

The miRNA loaded in the miRISC can recognize and bind target mRNAs, and gene silencing is then induced by cleavage or translational repression of mRNA transcripts, depending on the level of complementarity [17]. The mechanisms responsible for translation inhibition are not fully understood but may include blockage of translation initiation, dissociation of active ribosome or removal of cap/poly(A) tail of the mRNA [17, 18].

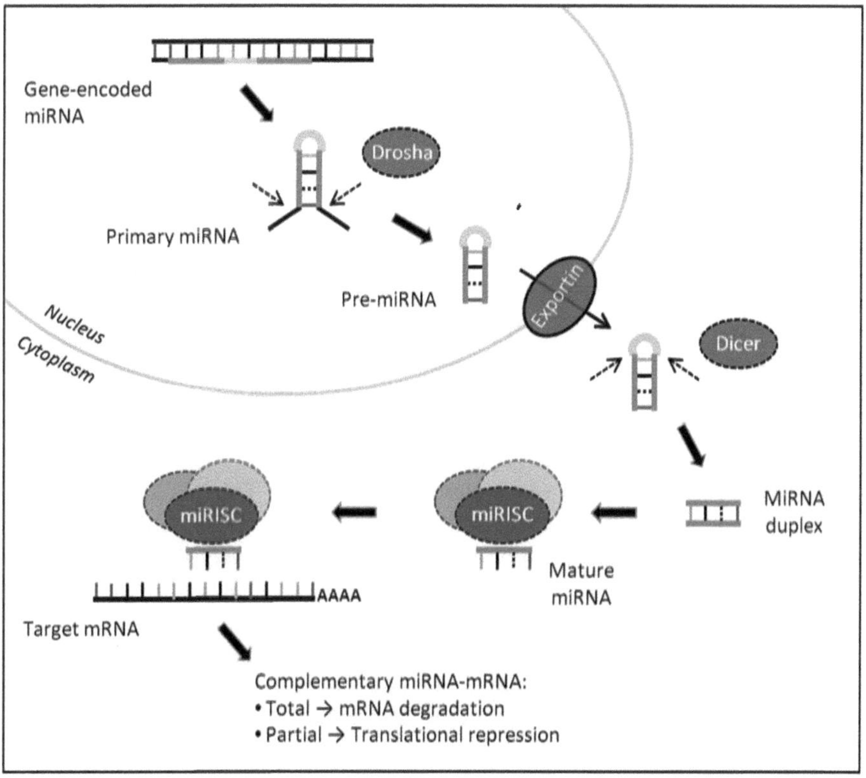

Fig. 9.1 Schematic representation of the microRNA (miRNA) biosynthesis pathway. The gene-encoded miRNA is transcribed to the primary miRNA (pri-miRNA), a long poly-adenylated mRNA containing a hairpin-loop (duplex structure). Then, the pri-miRNA is processed by Drosha (RNAse enzyme) to a 60–70 nucleotide pre-miRNA. This pre-miRNA is exported to the cytoplasm and processed by Dicer (endonuclease), resulting in a small double-stranded miRNA. The mature miRNA (antisense strand complementary to the target mRNA sequence) is loaded to the miRNA-Induced Silencing Complex (miRISC). This miRNA recognizes and binds target mRNAs, and gene silencing is induced by cleavage or translational repression of mRNA transcripts, depending on the level of complementarity

MicroRNAs and Obesity

Adipose Tissue and Adipogenesis

Excessive caloric intake promotes adipocyte hyperplasia and adiposity, which are hallmarks of obesity. Adipogenesis is the biological process that leads to the increase of adipocyte number due to the conversion of mesenchymal stem cells into preadipocytes, and their differentiation into adipocytes [19]. Mature adipocytes accumulate triglycerides and express specific genes such as peroxisome proliferator activator receptor gamma (PPAR-γ), CCAAT/enhancer binding protein alpha (C/

EBPα), GLUT4 and adipocyte protein homologous to myelin P2 (aP2), among others. The synchronized series of events that occur during adipogenesis are controlled by transcription factors and cell-cycle proteins regulating gene expression [19].

Role of microRNAs in Adipogenesis and Obesity

During the past years, numerous studies have shown that miRNAs participate in the physiological regulation of adipogenesis [20–25], and that obesity and high-fat diet can affect this regulation [26–29] (Tables 9.1 and 9.2). In fact, experiments with knocking-down essential enzyme for miRNA biogenesis confirm the importance of miRNAs in adipogenesis [20, 23, 30]. For example, suppression of Drosha or Dicer abolishes adipocyte differentiation [20, 30], while knock-down of Ago2 reduces the metabolic activity of adipocytes [23].

Research performed with mouse adipocyte cell line 3T3-L1, human primary adipocytes and mesenchymal stem cells showed that numerous miRNAs are differentially expressed during adipocytes maturation, suggesting that these particular miRNAs may participate in the cellular regulation of the adipogenic process [20–24, 27, 31–33]

Moreover, obesity leads to a deregulation of the miRNA pathway in the adipose tissue, as shown by the differential expression pattern of specific miRNAs in fat deposits of obese mice [27, 34–37] and obese subjects [31, 38–40] when compared to lean controls. Importantly, in obese and overweight humans, miRNAs deregulation in the adipose tissue co-relates with metabolic parameters such as body weight increase, body fat content, fasting plasma glucose, circulating levels of adipokines (leptin and adiponectin) and adipose tissue morphology [31, 38–40]. Together, these evidences support the role of miRNAs in the pathological development of obesity via impairment of adipocyte biology.

MicroRNAs that Promote Adipogenesis

Among the miRNAs that promote adipogenesis (Table 9.1), ***miR-143*** is one of the most investigated miRNAs in adipose tissue biology with several studies corroborating a pro-adipogenic role in mice and human adipocytes [21, 22, 27]. In fact, the expression levels of miR-143 are strongly upregulated during 3T3-L1 pre-adipocyte differentiation and primary human pre-adipocyte differentiation [21, 22, 27]. In accordance, ectopic expression of miR-143 promotes triglycerides accumulation during early differentiation of 3T3-L1 pre-adipocytes [27]. Moreover, inhibition of miR-143 in human pre-adipocytes suppressed the adipocyte markers GLUT4, PPAR-γ and triglycerides accumulation [21]. The mRNA targets of miR-143 that contribute to its adipogenic effect are largely unknown. It was first proposed that miR-143 could target ERK5 and regulate the MAP kinase signaling pathway to

Table 9.1 MicroRNAs that promote adipogenesis

miRNA	System	Function	Validated direct targets	Levels of miRNA in obesity	References
miR-17/92	Mouse adipocyte cell line	↑ Adipogenesis	Rb2/p130	N/a	[30]
miR-21	Human mesenchymal stem cells	↑ Adipogenesis Regulates proliferation of adipocytes precursors	Transforming growth factor β receptor 2 (TGFβR2) Signal transducer and activator of transcription 3 (STAT3)	Biphasic change in adipocytes of diet-induced obese mice (↓ after 1 week HFD; ↑ after 1 week HFD)	[34, 44]
miR-24	Mouse mesenchymal stem cell line	↑ Adipogenesis	N/A	N/A	[24]
miR-30c	Human mesenchymal stem cells	↑ Adipogenesis	Plasminogen activator inhibitor 1 (PAI1) Activin receptor-like kinase 2 (ALK2)	↓ In adipocytes of diet-induced obese mice ↓ In adipocytes of ob/ob obese mice	[35]
miR-30 family	Human mesenchymal stem cells	↑ Adipogenesis	Runt related transcription factor 2 (RUNX2)	N/A	[33]
miR-103	Mouse adipocyte cell line Mouse primary adipocytes	↑ Adipogenesis	N/A	↓ In adipocytes of diet-induced obese mice ↓ In adipocytes of ob/ob obese mice	[27]
miR-143	Mouse adipocyte cell line Mouse primary adipocytes Human primary adipocytes	↑ Adipogenesis	N/A	↓ In adipocytes of diet-induced obese mice (↑ was also reported) ↓ In adipocytes of ob/ob obese mice	[21, 22, 27, 42, 43]
miR-200c/141	Mouse mesenchymal stem cell line	↑ Adipogenesis	N/A	N/A	[46]
miR-200b, a/429	Mouse mesenchymal stem cell line	↑ Adipogenesis	N/A	N/A	[46]

Table 9.1 (continued)

miRNA	System	Function	Validated direct targets	Levels of miRNA in obesity	References
miR-210	Mouse adipocyte cell line	↑ Adipogenesis	Tcf712	N/A	[32]
miR-335	Mouse adipocyte cell line	↑ Adipogenesis	N/A	↑ In adipocytes of ob/ob obese mice ↑ In adipocytes of db/db obese mice	[36]
miR-375	Mouse adipocyte cell line	↑ Adipogenesis	N/A	N/A	[45]
miR-378	Mouse mesenchymal stem cell line Mouse adipocyte cell line	↑ Adipogenesis	N/A	N/A	[23]
miR-519d	Human primary adipocytes	↑ Adipogenesis	Peroxisome proliferator-activated receptor α (PPARα)	↑ In adipose tissue of human obese subjects	[38]

miR microRNAs, *N/A* not available, *HFD* high fat diet, ↑ increased, ↓ decreased

maintain the adipocyte differentiated state [21]. Other transcripts including fibroblast growth factor 7 (Fgf7) were also indicated as putative targets of miR-143 [41]; however, further studies are needed to validate these candidates.

Alterations in the biosynthesis of miR-143 may be responsible for the impairment of the adipogenic process observed in obesity [27, 42]. In fact, the levels of miR-143 are downregulated in adipocytes of obesity models, including high-fat diet (HFD) fed mice and leptin deficient (ob/ob) mice [27, 43]. However, another study reports that miR-143 is upregulated and associated to adipocyte differentiation markers PPAR-γ and aP2 in the white adipose tissue of HFD fed mice [42]. These opposite results may be influenced by experimental issues such as source of adipose tissue analyzed because miRNA expression profiles are significantly different between fat depot [39, 40]. Noteworthy, mice deficient for the miR-143–145 cluster develop white adipose tissue similarly to wild type mice, upon HFD regime [43]. These data suggest that miR-143 is dispensable for adipose tissue formation and maintenance in vivo [41, 43]. In conclusion, miR-143 has a proven modulatory role in adipogenesis in vitro, but the correlation between miR-143 and obesity in vivo and the targets of miR-143 involved in adipocyte differentiation need further investigation.

Similarly to miR-143, pro-adipogenic ***miR-103*** is strongly induced during murine and human adipogenesis [21, 27], and promotes triglycerides accumulation and adipocyte specific gene expression [27]. Moreover, the levels of miR-103 are down-regulated in adipocytes of obese mice [27].

Another example, ***miR-30c*** is upregulated in differentiated adipocytes, it promotes adipogenesis and is inhibited in genetic and diet obesity models [35]. Of note, the inverse expression pattern of miR-143, miR-103 and miR-30c observed in differentiated adipocytes and obese tissue [27, 35] suggests that obesity leads to a loss of miRNAs that characterize mature and metabolically active adipocytes.

Other pro-adipogenic miRNAs may play a role in the development of obesity by increasing the mature adipocyte population [36, 44]. For instance, ***miR-519d*** is upregulated in the adipose tissue of human obese subjects and ectopic expression of miR-519d promotes adipogenesis [38]. Also, the expression of ***miR-335*** is upregulated in the adipose tissue of genetically obese mice (leptin and leptin receptor deficient mice) and is correlated to differentiation markers in adipocyte cell line [36].

Another example, ***miR-21*** promotes adipogenic differentiation of human adipose tissue-derived mesenchymal stem cells via inhibition of TGF-β, a growth factor known to block adipogenesis in vitro and in vivo [44]. Additionally, miR-21 controls the number of adipocytes in murine adipose tissue by regulating the proliferation of adipocyte precursors at the early phase of obesity and increasing adipocyte differentiation during the late stage [34].

Several miRNAs that are upregulated in mature adipocytes have proven to stimulate adipogenesis after ectopic expression [30, 32, 33]. Moreover, the identification of mRNA targets helps understanding the role of those miRNAs in adipogenesis [30, 32, 33]. For example, the ***miR-17–92 cluster*** accelerates adipogenesis by targeting and negatively regulating Rb2/p130, tumor-suppressor proteins that inhibit cell differentiation [30]. ***MiR-210*** induces adipocyte differentiation by targeting Tcf7l2, an activator of the Wnt/β-catenin signaling, which should be suppressed during adipogenesis [32]. Additionally, the **miR-30** family directs the differentiation of human adipose tissue-derived stem cells to adipocytes via inhibition of osteogenesis transcription factor RUNX2 [33].

Furthermore, it was demonstrated that several other miRNAs including ***miR-24***, ***miR-375***, ***miR-378***, ***miR-200c/141*** and ***miR-200b, a/429*** can accelerate the differentiation of mesenchymal stem cells and pre-adipocytes to mature adipocytes [23, 24, 45, 46].

Considering the involvement of miRNAs in adipogenesis and obesity, a novel strategy to reduce fat deposit based on the decrease of pro-adipogenic miRNA levels emerges; this strategy would aim an inhibition of pre-adipocyte expansion and/or differentiation. However, caution should be taken because this putative therapeutic approach may lead to toxic accumulation of lipids in the liver and other organs.

MicroRNAs that Inhibit Adipogenesis

MiRNAs can also regulate adipogenesis negatively as reported by recent murine and human studies [25, 26, 28, 29, 47, 48] (Table 9.2). For example, the miR-27 family (***miR-27a, b***) is down-regulated during differentiation of 3T3-L1 pre-adipocytes and human mesenchymal stem cells [26, 28, 29]. Moreover, miR-27a or miR-27b mimics specifically inhibit adipogenic differentiation of murine pre-adipocytes and mesenchymal stem cells and block the expression of PPARγ and C/EBPα, two master regulators of adipogenesis, when delivered before adipogenesis induction [26, 28, 29]. In fact, the PPARγ mRNA transcript was validated as target of miR-27b and miR-27a [28, 29]. Interestingly, the anti-adipogenic effects of miR-27a, b are not observed when mimics are delivered few days after adipogenesis induction [26]. These data suggest that miR-27 family inhibits adipogenesis by preventing pre-adipocyte from entering the differentiation process.

The expression of ***miR-27a, b*** is increased in fat tissue of genetically obese ob/ob mice and regulated by hypoxia, an important extracellular stress associated with obesity [26]. However, the levels of miR-27a are lower in the adipose tissue of dietary obese mice when compared to lean controls [28]. Despite these discrepancies and knowing the important role of miR-27 family in the regulation of adipogenesis, alterations in the expression levels of these miRNAs may be contributing to an inappropriate adipocyte differentiation in obesity.

The ***miRNA let-7*** is well known to regulate cell proliferation and differentiation processes of different cell types [49]. In what concerns adipocytes, let-7 inhibits the expansion and terminal maturation of 3T3-L1 pre-adipocytes [47]. The levels of let-7 increase during adipogenesis and in differentiated mouse adipocytes, and this miRNA seems to act via negative regulation of HMGA2 [47]. HMGA2 is a transcription factor that controls growth and proliferation in other contexts and is necessary for adipose tissue formation in transgenic mice models [50]. In accordance, transgenic mice with global overexpression of let-7 show decreased body adiposity as they fail to accumulate fat with age [51], indicating a putative role of let-7 in the expansion of adipose tissue in vivo.

MiR-31 has emerged as key regulator of adipogenesis because it targets C/EBPα, the major factor promoting mesenchymal stem cell differentiation to adipocytes [24]. In fact, miR-31 overexpression prevents the upregulation of C/EBPα during adipogenesis induction of murine mesenchymal stem cell line [24]. This information suggests that miR-31 has an anti-adipogenic effect by preventing cell commitment towards adipocyte maturation.

Other miRNAs have been also identified as negative regulators of mesenchymal stem cell differentiation to adipocytes [25, 48]. For instance, ***miR-363*** is significantly down-regulated during the adipogenic process, and overexpression of miR-363 inhibits expansion and terminal differentiation of mesenchymal stem cells [25]; suggesting that miR-363 levels should be decreased to allow the maturation of adipocytes. The anti-adipogenic effect is mediated by direct inhibition of E2F3 mRNA transcripts, a key transcription factor that regulates growth and proliferation [25]. Another example is ***miR-138***, which inhibits adipocyte differentiation and

Table 9.2 MicroRNAs that inhibit adipogenesis

miRNA	System	Function	Validated direct targets	Levels of miRNA in obesity	References
Let-7	Mouse adipocyte cell line	↓ Adipogenesis	High mobility group protein A (HMGA2)	N/A	[47]
miR-15	Mouse adipocyte cell line	↓ Adipogenesis	Delta-like 1 homolog (DLK1)	N/A	[53]
miR-27a/b	Human mesenchymal stem cells Mouse adipocyte cell line	↓ Adipogenesis	Peroxisome proliferator activator receptor γ (PPARγ)	↓ In adipocytes of diet-induced obese mice ↑ In adipocytes of ob/ob obese mice	[26, 28, 29]
miR-31	Mouse mesenchymal stem cell line	↓ Adipogenesis	CCAAT/enhancer binding protein α (C/EBPα)	N/A	[24]
miR-138	Human mesenchymal stem cells	↓ Adipogenesis	Early region 1-A-like inhibitor of differentiation 1 (EID1)	N/A	[48]
miR-363	Mouse mesenchymal stem cell line	↓ Adipogenesis	E2F3	N/A	[25]
miR-448	Mouse adipocyte cell line	↓ Adipogenesis	Krüppel-like factor 5 (KLF5)	N/A	[52]

miR microRNAs, *N/A* not available, ↑ increased, ↓ decreased

decreases the expression levels of key adipogenic markers, via targeting the nuclear receptor EID-1 (Early region 1-A-like inhibitor of differentiation 1) [48].

Additionally, ***miR-448*** inhibits adipogenesis when transfected to 3T3-L1 pre-adipocytes before induction of differentiation [52]. MiR-448 exerts its anti-adipogenic actions, at least partially, by targeting the transcription factor KLF5 (Krüppel-like factor 5) [52]. ***MiR-15a*** seems to be involved in the determination of 3T3-L1 pre-adipocytes cell number and cell size via DLK1, a molecule previously implicated in cell growth [53].

During the last years, several potential negative regulators of adipogenesis were identified but most of the research was performed in vitro. The implications of these miRNAs in adipogenesis in vivo need further clarification before conclusions can be drawn for their role in adipose tissue dysfunction in obesity.

Extracellular Factors Regulate miRNAs in Adipocytes

The adipose tissue of obese mammals and humans is exposed to a chronic inflammatory environment caused by macrophages infiltration [54]. In this situation, the elevated levels of inflammatory cytokines, such as TNF-α and IL6, are responsible for the metabolic adipocyte dysfunctions including insulin resistance [55].

Inflammatory mediators can also induce miRNA alterations similar to the ones observed in obesity [27, 56]. In fact, 3T3-L1 adipocytes treated with TNF-α show reduced expression of the same miRNAs that are down-regulated in the adipose tissue of ob/ob mice, including miR-103 and miR-143 [27]; and increased expression of miRNAs that are upregulated in the adipose tissue of dietary and genetic obese mice [27]. The levels of cytokine transforming growth factor beta (TGF-β) and miR-130b are increased in the adipose tissue of obese ob/ob mice [56]. Interestingly, incubation of mouse adipocytes with TGF- β elevates the quantity of miR-130b secreted by these cells, mimicking the obesity-induced alteration of miR-130b [56]. These studies show that miRNA alterations in obesity are, at least partially, caused by elevated levels of inflammation mediators, and are in line with the notion that miRNAs are associated to the metabolic dysfunction of obese adipose tissue. Nevertheless, further studies are needed to establish the mechanistic link between cytokines and adipocyte miRNA expression, and to unravel the putative role of other inflammatory mediators in this process.

Another interesting concept is the regulation of miRNA levels in adipocytes by dietary components. In fact, diet supplementation with conjugated linoleic acid (CLA) affects the expression of miRNAs in murine adipose tissue, in a dose-dependent manner [57]. Moreover, the miRNAs altered by dietary CLA were previously implicated in adipogenesis and adipocyte alterations induced by obesity, including miR-143, miR-107, miR-221 and miR-222 [27, 57]. This investigation introduces the possibility that dietary components can affect fat deposits via miRNA pathway. In the future, it will be critical to understand whether dietary lipids present in western diets can change the miRNA profile in human adipose tissue.

Glucose Metabolism

Obesity and high-fat diet are frequently associated to type 2 diabetes (T2D) and glucose metabolism impairment, which are characterized by deficient insulin production from the pancreas and insulin resistance from target tissues. Over the last years, miRNAs have emerged as regulators of glucose homeostasis by influencing cellular mechanisms that control insulin production and secretion [58–61], insulin resistance and glucose sensitivity [43, 62–64] (Table 9.3). Moreover, specific miRs were associated to T2D and obesity in rodent models and human subjects [58, 65–68].

Table 9.3 MicroRNAs involved in glucose metabolism

miRNA	System	Function	Validated direct targets	Levels of miRNA in obesity	References
Let-7	Mouse pancreas Mouse liver and muscle	↓ Insulin secretion ↑ Insulin resistance	Insulin receptor (IR) Insulin receptor substrate 2 (IRS2) AKT-2 Insulin-like growth factor 1 receptor (IGF1R) PIK3 interacting protein 1	Silencing let-7 protects and treats obesity-induced insulin resistance	[51, 62]
miR-9	Rodent pancreatic islets Mouse pancreatic β-cell line	↓ Insulin secretion	Onecut-2 (OC2) Sirtuin-1 (SIRT1)	↑ In the presence of high glucose	[59, 72]
miR-15a	Mouse pancreatic β-cell line	↑ Insulin expression	Uncoupling protein-2 (UCP2)	Modulated by glucose levels	[69]
miR-24	Pancreatic β-cells	↓ Insulin secretion	HNF1A NEUROD1	↑ In pancreatic islets of obese db/db mice ↑ In pancreatic islets of diet-induced obese mice	[67]
miR-29	Adipocyte cell line Hepatic cell line	↓ Glucose up-take (adipocytes) ↓ Insulin signaling (hepatocytes)	PIK3 subunit p85α	↑ In muscle, fat and liver of diabetic rats ↑ In liver of db/db mice ↑ In the presence of high glucose or insulin (adipocytes)	[63, 81]
miR-30d	Mouse pancreatic β-cell line	↑ Insulin expression	N/A	↑ In the presence of high glucose	[61]
miR-33a/b	Human hepatic cell line	↓ Insulin signaling ↓ glucose uptake	IRS2 SIRT6	N/A	[84]

Table 9.3 (continued)

miRNA	System	Function	Validated direct targets	Levels of miRNA in obesity	References
miR-34a	Mouse pancreatic β-cell line Rat pancreatic islets	↓ Insulin secretion	Vesicle-associated membrane protein 2 (VAMP2) BCL-2	↑ β-cells exposed to fatty acid palmitate ↑ In pancreatic islets of diabetic db/db mice	[66]
miR-96	Mouse pancreatic β-cell line	↓ Insulin secretion	N/A	N/A	[71]
miR-103/ miR-107	Mouse liver and adipose tissue	↑ Insulin resistance ↓ Glucose uptake (adipocytes)	Caveolin1 (CAV1)	↑ In liver of diet-induced obese mice ↑ In liver of obese ob/ob mice	[64]
miR-126	Hepatic cell line	↑ Insulin resistance ↓ Insulin signaling	IRS1	↑ In hepatocytes with mitochondrial dysfunction	[83]
miR-143/ miR-145	Mouse liver	↑ Insulin resistance	Oxysterol binding protein-like 8 (ORP8)	↑ In liver of diet-induced obese mice ↑ In liver of obese db/db mice	[43]
miR-144	T2D rats and humans Rat pancreatic islets	Insulin signaling	IRS1	↑ In pancreas, adipose tissue and liver of T2D rats ↑ In the presence of high glucose	[58]
miR-146	Mouse pancreatic β-cell line Rat pancreatic islets	N/A	N/A	↑ β-cells exposed to fatty acid palmitate ↑ In pancreatic islets of diabetic db/db mice	[66]

Table 9.3 (continued)

miRNA	System	Function	Validated direct targets	Levels of miRNA in obesity	References
miR-375	Mouse pancreatic β-cell line Rat pancreatic islets	↓ Insulin secretion ↓ Insulin expression	3 phosphoinositide dependent protein kinase-1 (PDK1) Myotrophin (MTPN)	↑ In pancreas of obese ob/ob mice ↓ In pancreas of diabetic rats ↓ In the presence of high glucose	[60, 71, 75]
miR-802	Mouse hepatic cell line	↓ Insulin signaling ↑ Insulin resistance	Hepatocyte nuclear factor 1β (HNF1β)	↑ In liver of diet-induced obese mice ↑ In liver of obese db/db mice (also in adipose tissue, pancreatic islets and muscle) ↑ In liver of overweight humans	[82]

miR microRNAs, *N/A* not available, *T2D* type 2 diabetes, ↑ increased, ↓ decreased

Insulin Production and Secretion

Some studies highlighted the role of miRNAs on insulin production and secretion by β-cells [58–61], and, consequently, in the establishment of T2D. Indeed, profiling studies show that several miRs are differentially expressed in the pancreatic islets of T2D rat models [58, 65–67] and obese mice [67, 68]. Moreover, glucose regulates the expression levels of these differentially expressed miRNAs in T2D rat pancreas, and miRNAs involved in insulin biosynthesis in pancreatic cell lines and human islets cultures [58, 61, 65, 69, 70]. These evidences suggest a regulatory loop between circulating glucose levels and insulin production that involves the miRNA pathway.

Concerning insulin secretion, ***miR-124a***, ***miR-96*** and ***miR-9*** were identified as negative regulators of glucose-stimulated insulin release in pancreatic cell lines and rodent pancreatic islets [59, 71, 72]. These miRNAs act by affecting the expression of several components of the exocytotic machinery, as well as transcription factors involved in intracellular signaling [59, 71–73]. Other miRNAs, including ***miR-30d*** and ***miR-15a*** were implicated as positive regulators of insulin gene expression, but not insulin release, in murine pancreatic β-cell line MIN6 [61, 69].

MiR-144 may also have an important role in insulin metabolism because it is upregulated in pancreatic islets of diabetic rats, and it targets the insulin receptor substrate 1 (IRS1) [58]. IRS1 present in pancreatic β-cells promotes insulin secretion in an autocrine manner [74]. Inhibition of IRS1 by miR-144 may contribute to insulin secretion impairment in diabetes.

The islets specific ***miR-375*** suppresses glucose-induced insulin secretion from murine pancreatic β-cell line [60, 71]. MiR-375 knock-out mice exhibit severe defects in numerous glucose metabolism markers, further supporting the importance of miR-375 as a regulator of insulin-dependent mechanisms [68]. MiR-375 directly targets and reduces PDK1 (3-phosphoinositide-dependent protein kinase-1), a protein involved in the phosphatidylinositol (PI) 3-kinase cascade [75]; this is relevant in pancreatic cells because glucose stimulates the insulin gene promoter activity via the PI 3-kinase cascade. Noteworthy, obese ob/ob mice have increased expression of miR-375 in the pancreas [68], which may represent a link between obesity and insulin secretion deficiency.

MiR-34a and ***miR-146*** affect exocytosis and, consequently, the secretion of insulin from β-cells [66]. These miRNAs are upregulated in the pancreatic islets of diabetic db/db mice and in β-cells exposed to fatty acid palmitate [66]. Plasma fatty acids are commonly elevated in high-fat diet regimes and produce detrimental consequences in pancreatic β-cells; as shown in this study, this adverse effect is, at least partially, mediated by alterations in the levels of specific miRNAs [66].

A recent study showed that ***miR-24*** is highly expressed in pancreatic β-cells and further upregulated in islets from genetic fat mice (db/db) or mice fed with a high-fat diet [67]. Additionally, overexpression of miR-24 inhibits insulin secretion [67]. In the future, it will be relevant to evaluate the levels of other pancreatic β-cells miRNAs in rodent models of obesity and obese subjects and further investigate a putative role of miRNA dysfunction in the impairment of insulin secretion associated to obesity. Delivery of specific miRNAs to the pancreatic islets may represent a putative therapy to reestablish insulin secretion in patients that developed insulin deficiency owing to high-fat diet regime and obesity.

Glucose Sensitivity and Insulin Resistance

Insulin binding to the insulin receptor (IR) activates intracellular signaling pathways that include the proteins insulin receptor substrate (IRS), PI3K, PDK1 and the protein kinase AKT, resulting in glucose uptake from target cells [76–78]. Deregulation of this process is responsible for insulin resistance (diminished glucose uptake and utilization in insulin-sensitive tissues), which is one the most frequent alterations observed in T2D obese subjects and murine obese models [76–78].

A growing body of evidences indicates that miRNAs participate in the development of insulin resistance as suggested by the differential expression of specific miRNAs in the muscle, liver and adipose tissue of T2D rat models [58, 79–81] and obese mice [43, 64, 82]. In this situation, alterations of miRNA expression levels can affect gene expression of transcripts involved in insulin signaling in target tissues.

The ***miR-103/107*** family is upregulated in the liver of obese ob/ob mice and diet-induced obese mice [64]. Moreover, silencing of miR-103/107 leads to improved glucose tolerance and insulin sensitivity in the liver and adipose tissue of those mice [64], which implicates these miRNAs as possible therapeutic targets for diabetes. Conversely, miR-103/107 gain of function in the liver or fat is sufficient to induce insulin resistance and glucose intolerance in mice fed with standard chow diet, as shown in glucose tolerance and insulin tolerance tests [64]. Caveolin-1 (Cav1) is a direct target of miR-103/107; this protein regulates the number of insulin receptors in cell membranes [64]. Therefore, elevated expression of miR-103/107 in obese mice will decrease the levels of Cav1, reduce insulin signaling and contribute to insulin resistance.

Another miRNA involved in hepatocyte insulin signaling is ***miR-802*** [82]. This miRNA is upregulated in the liver of obese mice and humans and induces insulin resistance by targeting the hepatocyte nuclear factor 1β (HNF1β) [82]. Transgenic mice overexpressing miR-802 have impaired insulin homeostasis including insulin insensitivity and glucose intolerance [82].

The expression levels of ***miR-143/miR-145*** cluster are increased in the liver of dietary and genetic db/db obese mice [43]. This alteration is implicated in insulin insensitivity since inducible hepatic miR-143 overexpression results in prominent deregulation of glucose homeostasis [43]. In accordance, mice deficient for miR-143/145 are protected from diet-induced insulin resistance [43]. MiR-143 is a negative regulator of insulin sensitivity by indirectly inhibiting the ability of insulin to phosphorylate and activate AKT and downstream kinase signaling in hepatic cells [43].

MiR-29 also inhibits the phosphorylation of AKT, in adipocyte and hepatocyte cell lines, but without altering the total levels of this insulin signaling cascade kinase [63, 81]. The indirect effect of miR-29 in AKT phosphorylation is mediated by negative regulation of upstream mediator PIK3 subunit p85α [63]. MiR-29 seems to be responsible for insulin resistance in adipocytes because it inhibits insulin-induced glucose uptake when delivered to these cells [81]. Further supporting this hypothesis, the expression of miR-29 is elevated in insulin target tissues (muscle, fat, liver) of diabetic rats and diabetic db/db mice [63, 81].

Mitochondrial dysfunction is associated to the development of insulin resistance in T2D, and, more recently, miRNAs were implicated in this process [83]. Indeed, ***miR-126,*** which is upregulated in hepatocytes with mitochondrial dysfunction, regulates negatively the IRS1 mRNA transcript, thus inhibiting intracellular insulin signaling [83]. Similarly, the IRS2 transcript is a target of ***miR-33a/b*** in hepatocytes [84]. Overexpression of miR-33a/b results in an insulin resistance phenotype that includes decreased insulin signaling and glucose uptake [84].

Recently, the ***Lin-28/let-7*** axis was implicated in glucose metabolism in mice [51, 62]. The let-7 family consists of 9 slightly different miRNAs that share the same seed region (let-7a, let7-b, let7-c, let7-d, let-7e, let7-f, let7-g, let7-i and miR-98) [85]; the let-7 biogenesis is inhibited by the RNA-binding proteins Lin-28a and Lin-28b [85].

Global let-7 transgenic mice present body composition alterations namely reduced fat mass and body weight gain with age [51]. Let-7 transgenic mice also exhibit glucose metabolism alterations, including high levels of glucose in glucose-tolerance test [51, 62]. However, it remains to be clarified whether the glucose homeostasis defects result from insufficient insulin secretion [51] or insulin resistance from peripheral tissues, that was associated to compensatory overproduction of insulin [62]. The discrepancies observed in these studies could be related to the different overexpression pattern of let-7 in the two transgenic mice models, in particular in what concerns the pancreas [51, 62]. Indeed, mice with specific overexpression of let-7 in the pancreas (but not in the muscle, liver, adipose tissue or neurons) exhibit decreased insulin secretion [51].

In the context of dietary obesity, global knockdown of let-7 using anti-miR molecules is sufficient to prevent and treat glucose intolerance, presumably trough improved insulin sensitivity in peripheral tissues [51]. The anti-miR let-7 treatment could also reduce fat accumulation and increase lean mass in mice fed with high-fat diet [51]. Moreover, this phenotype is replicated in transgenic mice overexpressing the let-7 biosynthesis inhibitors Lin28a/b [62]. Nevertheless, the involvement of let-7 in the pathogenesis of diet-induced obesity and diabetes is still unclear because high-fat diet did not change the levels of let-7 in metabolic organs [51].

Knowing that let-7 induces insulin resistance, it was investigated whether insulin signaling mediators are targeted by let-7 for negative regulation [62]. Indeed, in vitro studies confirmed that let-7 targets the mRNA transcripts of IR and IRS-2 and reduces the expression levels of these proteins [62]. Moreover, anti-miR let-7 treatment prevents the down-regulation of IR and IRS-2 in the muscle and liver of obese mice, which may contribute to the positive effects that this possible anti-obesity therapy has on glucose metabolism [51]. Other transcripts of the insulin signaling pathway were validated as targets of let-7, namely AKT-2, insulin-like growth factor 1 receptor and PIK3 interacting protein 1 [62].

Despite the functional evidence for a role of altered posttranscriptional gene silencing in the development of obesity-associated insulin resistance in vivo, more investigation is needed in this area. For example, it will be important to define the molecular mechanisms leading to alteration of miRNA expression levels in insulin target tissues in obesity. Nevertheless, the experiments showing improvement of glucose homeostasis in obesity mice models after silencing of specific miRs implicated in insulin insensitivity [43, 51, 64], uncover a therapeutic option that consists in using anti-miRs to reverse obesity-associated insulin resistance. In particular, the anti-let-7 treatment presents a high therapeutic potential for dietary obesity because of the combined positive effects on insulin secretion, prevention of glucose intolerance and reduction of fat deposits [51].

This putative therapeutic strategy may be more effective if based on the modulation of miRNA clusters or families, rather than a single miRNA. Indeed, as suggested from the roles of miR-103/107, miR-143/145 and let-7 family in insulin signaling, a hypothesis arise that different members of one miRNA cluster/family can act together to coordinately control a signal transduction pathway at multiple levels. This theory was reinforced by Xu and collaborators, in 2008, showing that each member of the miRNA cluster mir-183/96/182 targets one mediator of the insulin signaling pathway [86].

Circulating microRNAs

MiRNAs can be detected in human blood and biofluids; they circulate inside microvesicles or exosomes that are derived from the plasmatic membranes of donor cells [87] or as lipid-free complexes with AGO2 protein [88]. Circulating miRNAs act as endocrine or paracrine signaling molecules, delivering regulatory messages from donor cells to target cells or tissues [89, 90]. The miRNA-mediated cross-talk between metabolic tissues was investigated more recently in an interesting study that suggested that serum miR-130b high levels observed in obese mice are related to increased miRNA secretion from mature adipocytes [56]. Moreover, the miR-130b levels are also increased in the muscle but the transcription activation for this miRNA occurs only in adipocytes, indicating that miR-130b secreted by fat cells is uptaken by muscle cells [56].

Another interesting concept is the fact that aberrant levels of circulating miRs indicate a pathological condition from donor cells [91]. In what concerns obesity, evidences indicate that the levels of blood miRNAs reflect their expression in metabolic organs [40, 56, 58]. For example, a study showed that miRNAs down-regulated in the adipose tissue of obese patients (miR-17 and miR-132) are also decreased in the blood of these individuals [40]. Additionally, a significant number of miRNAs that are altered in metabolic organs of diet-induced obese rats are simultaneously deregulated in their blood samples [58].

Hence, circulating miRNAs have been investigated as possible non-invasive disease biomarkers. Moreover, miRNAs are expressed in a tissue specific way and associated with pathological mechanism, which reinforces their ability to serve as biomarkers. Indeed, altered plasma or serum miRNA profiles have been linked to numerous diseases including cancer, cardiac disease, hepatic injury and infection [92–96]. In addition, studies have also reported alterations of specific circulating miRNAs in T2D and obese patients [40, 56, 58, 93, 97].

In line with these observations, it was shown that T2D patients, pre-diabetic patients and control subjects have distinct profiles of circulating miRNAs [58, 97]. Therefore, the metabolic condition of an individual (diabetic or pre-diabetic) can be distinguished by determining the blood levels of specific miRNAs [58]. Besides, the levels of particular blood miRNAs correlate with obesity indicators in humans, including body mass index, body fat percentage and triglyceride levels, and can be used in the diagnostic of individuals with metabolic syndrome [56]. Importantly, a study reported that the development of T2D by normoglycemic subjects could be predicted by a specific plasma miRNA profile (reduction of miR-15a, miR-29b, miR-126 and miR-223 and elevation of miR-28), within a 10-years period before disease manifestation [97]. Altogether, these evidences highlight the promising use of blood miRNA profiling in the identification of patients with increased risk of developing T2D.

The use of miRNAs as obesity or disease biomarkers presents numerous advantages namely, non-invasive collection method, good stability in plasma and serum samples and reproducible levels among individuals of the same specie [96]. However, there are still some issues that need to be addressed including validation of miRNA profiles in large studies, standardization of quantification methods and source of miRNAs (total blood or plasma/serum).

Conclusion and Future Perspectives

Recent investigation revealed the critical contribution of miRNAs to the regulation of cellular mechanisms involved in metabolism, including adipogenesis and glucose metabolism. Furthermore, deregulation of the miRNA pathway was implicated in the cellular alterations present in obese metabolic tissues. In this context, miRNA modulation emerges as a possible therapeutic option to reestablish the regulatory balance in biological pathways affect by obesity.

The unique characteristics of miRNAs designate them as interesting candidates for an effective therapy against obesity and metabolic diseases. For example, one single miRNA can target several metabolic pathways or multiple molecules acting at different levels of the same pathway. Theoretically, the one molecule with multiple-targets approach will be more efficient than the traditional pharmacological options focused in one particular receptor or protein. However, this is also one important limitation of using miRNAs as therapy because it may result in unspecific or adverse effects in other tissues or signaling pathways.

Modulation of specific miRNAs with therapeutic purposes can be achieved with synthetic nucleic acids that function as miRNA mimics or inhibitors. These molecules can be delivered to the organism by peripheral roots of administration and achieve miRNA overexpression or silencing in multiple organs [98]. For example, subcutaneous administration of locked-nucleic acids (LNA) anti-let-7 to obese mice decreases the expression of this miRNA in the liver, pancreas and fat tissue, up to 2 weeks after the last injection [51]. Moreover, this strategy as proven to prevent and treat glucose intolerance in mouse models of obesity [51]. However, repeated administrations and high doses are required to maintain the effect of synthetic nucleic acids.

On the other hand, most of the promising results concerning the use of LNA in metabolic diseases have been achieved by silencing individual miRNAs in the liver, as this organ seems to be more accessible to nucleic acids [41, 82]. Indeed, intravenous administration of LNA anti-miR-802 specifically reduces the hepatic levels of this miRNA and improves insulin sensitivity in dietary obese mice [82]. In accordance, higher doses of anti-miR are needed to achieve miR-143 silencing in the adipose tissue as compared to the liver [41].

Further investigation is needed to clarify the distribution pattern of synthetic nucleic acids in the living organism and investigate alternative approaches to allow miRNA modulation specifically in other tissues affected by obesity. In the context of metabolism, organ specific modulation of miRNAs was achieved by different approaches including lipid modifications of synthetic nucleic acid molecules, lipid based formulations and viral vectors [64].

In the future, the miRNA biology and the function of these regulatory molecules in different metabolic pathways will be topics of intense investigation, aiming a better understanding of the cellular mechanism affected in obesity and a possible new innovative therapy.

References

1. Bartel DP (2004) MicroRNAs: genomics, biogenesis, mechanism, and function. Cell 116(2):281–297
2. Lagos-Quintana M, Rauhut R, Lendeckel W, Tuschl T (2001) Identification of novel genes coding for small expressed RNAs. Science 294(5543):853–858
3. Guo H, Ingolia NT, Weissman JS, Bartel DP (2010) Mammalian microRNAs predominantly act to decrease target mRNA levels. Nature 466(7308):835–840
4. Lee RC, Feinbaum RL, Ambros V (1993) The C. elegans heterochronic gene lin-4 encodes small RNAs with antisense complementarity to lin-14. Cell 75(5):843–854
5. Wightman B, Ha I, Ruvkun G (1993) Posttranscriptional regulation of the heterochronic gene lin-14 by lin-4 mediates temporal pattern formation in C. elegans. Cell 75(5):855–862
6. Rottiers V, Naar AM (2012) MicroRNAs in metabolism and metabolic disorders. Nat Rev Mol Cell Biol 13(4):239–250
7. Bartel DP (2009) MicroRNAs: target recognition and regulatory functions. Cell 136(2):215–233
8. Doench JG, Sharp PA (2004) Specificity of microRNA target selection in translational repression. Genes Dev 18(5):504–511
9. Baek D, Villén J, Shin C, Camargo FD, Gygi SP, Bartel DP (2008) The impact of microRNAs on protein output. Nature 455(7209):64–71
10. Cai X, Hagedorn CH, Cullen BR (2004) Human microRNAs are processed from capped, polyadenylated transcripts that can also function as mRNAs. RNA 10(12):1957–1966
11. Lee Y, Kim M, Han J, Yeom KH, Lee S, Baek SH, Kim VN (2004) MicroRNA genes are transcribed by RNA polymerase II. EMBO J 23(20):4051–4060
12. Gregory RI, Yan KP, Amuthan G, Chendrimada T, Doratotaj B, Cooch N, Shiekhattar R (2004) The Microprocessor complex mediates the genesis of microRNAs. Nature 432(7014):235–240
13. Lee Y, Ahn C, Han J, Choi H, Kim J, Yim J, Lee J, Provost P, Rådmark O, Kim S, Kim VN (2003) The nuclear RNase III Drosha initiates microRNA processing. Nature 425(6956):415–419
14. Haase AD, Jaskiewicz L, Zhang H, Lainé S, Sack R, Gatignol A, Filipowicz W (2005) TRBP, a regulator of cellular PKR and HIV-1 virus expression, interacts with Dicer and functions in RNA silencing. EMBO Rep 6(10):961–967
15. Hutvagner G, McLachlan J, Pasquinelli AE, Bálint E, Tuschl T, Zamore PD (2001) A cellular function for the RNA-interference enzyme Dicer in the maturation of the let-7 small temporal RNA. Science 293(5531):834–838
16. Chendrimada TP, Gregory RI, Kumaraswamy E, Norman J, Cooch N, Nishikura K, Shiekhattar R (2005) TRBP recruits the Dicer complex to Ago2 for microRNA processing and gene silencing. Nature 436(7051):740–744
17. Huntzinger E, Izaurralde E (2011) Gene silencing by microRNAs: contributions of translational repression and mRNA decay. Nat Rev Genet 12(2):99–110

18. Ku G, McManus MT (2008) Behind the scenes of a small RNA gene-silencing pathway. Hum Gene Ther 19(1):17–26
19. Tang QQ, Lane MD (2012) Adipogenesis: from stem cell to adipocyte. Annu Rev Biochem 81:715–736
20. Oskowitz AZ, Lu J, Penfornis P, Ylostalo J, McBride J, Flemington EK, Prockop DJ, Pochampally R (2008) Human multipotent stromal cells from bone marrow and microRNA: regulation of differentiation and leukemia inhibitory factor expression. Proc Natl Acad Sci U S A 105(47):18372–18377
21. Esau C, Kang X, Peralta E, Hanson E, Marcusson EG, Ravichandran LV, Sun Y et al (2004) MicroRNA-143 regulates adipocyte differentiation. J Biol Chem 279(50):52361–52365
22. Kajimoto K, Naraba H, Iwai N (2006) MicroRNA and 3T3-L1 pre-adipocyte differentiation. RNA 12(9):1626–1632
23. Gerin I, Bommer GT, McCoin CS, Sousa KM, Krishnan V, MacDougald OA (2010) Roles for miRNA-378/378* in adipocyte gene expression and lipogenesis. Am J Physiol Endocrinol Metab 299(2):E198–206
24. Sun F, Wang J, Pan Q, Yu Y, Zhang Y, Wan Y, Wang J, Li X, Hong A (2009) Characterization of function and regulation of miR-24-1 and miR-31. Biochem Biophys Res Commun 380(3):660–665
25. Chen L, Cui J, Hou J, Long J, Li C, Liu L (2014) A novel negative regulator of adipogenesis: microRNA-363. Stem Cells 32(2):510–520
26. Lin Q, Gao Z, Alarcon RM, Ye J, Yun Z (2009) A role of miR-27 in the regulation of adipogenesis. FEBS J 276(8):2348–2358
27. Xie H, Lim B, Lodish HF (2009) MicroRNAs induced during adipogenesis that accelerate fat cell development are downregulated in obesity. Diabetes 58(5):1050–1057
28. Kim SY, Kim AY, Lee HW, Son YH, Lee GY, Lee JW, Lee YS, Kim JB (2010) miR-27a is a negative regulator of adipocyte differentiation via suppressing PPARgamma expression. Biochem Biophys Res Commun 392(3):323–328
29. Karbiener M, Fischer C, Nowitsch S, Opriessnig P, Papak C, Ailhaud G, Dani C, Amri EZ, Scheideler M (2009) microRNA miR-27b impairs human adipocyte differentiation and targets PPARgamma. Biochem Biophys Res Commun 390(2):247–251
30. Wang Q, Li YC, Wang J, Kong J, Qi Y, Quigg RJ, Li X (2008) miR-17-92 cluster accelerates adipocyte differentiation by negatively regulating tumor-suppressor Rb2/p130. Proc Natl Acad Sci U S A 105(8):2889–2894
31. Ortega FJ, Moreno-Navarrete JM, Pardo G, Sabater M, Hummel M, Ferrer A, Rodriguez-Hermosa JI et al (2010) MiRNA expression profile of human subcutaneous adipose and during adipocyte differentiation. PLoS ONE 5(2):e9022
32. Qin L, Chen Y, Niu Y, Chen W, Wang Q, Xiao S, Li A et al (2010) A deep investigation into the adipogenesis mechanism: profile of microRNAs regulating adipogenesis by modulating the canonical Wnt/beta-catenin signaling pathway. BMC Genomics 11:e320
33. Zaragosi LE, Wdziekonski B, Brigand KL, Villageois P, Mari B, Waldmann R, Dani C, Barbry P (2011) Small RNA sequencing reveals miR-642a-3p as a novel adipocyte-specific microRNA and miR-30 as a key regulator of human adipogenesis. Genome Biol 12(7):R64
34. Kim YJ, Hwang SH, Cho HH, Shin KK, Bae YC, Jung JS (2012) MicroRNA 21 regulates the proliferation of human adipose tissue-derived mesenchymal stem cells and high-fat diet-induced obesity alters microRNA 21 expression in white adipose tissues. J Cell Physiol 227(1):183–193
35. Karbiener M, Neuhold C, Opriessnig P, Prokesch A, Bogner-Strauss JG, Scheideler M (2011) MicroRNA-30c promotes human adipocyte differentiation and co-represses PAI-1 and ALK2. RNA Biol 8(5):850–860
36. Nakanishi N, Nakagawa Y, Tokushige N, Aoki N, Matsuzaka T, Ishii K, Yahagi N et al (2009) The up-regulation of microRNA-335 is associated with lipid metabolism in liver and white adipose tissue of genetically obese mice. Biochem Biophys Res Commun 385(4):492–496

37. Chartoumpekis DV, Zaravinos A, Ziros PG, Iskrenova RP, Psyrogiannis AI, Kyriazopoulou VE, Habeos IG (2012) Differential expression of microRNAs in adipose tissue after long-term high-fat diet-induced obesity in mice. PLoS ONE 7(4):e34872
38. Martinelli R, Nardelli C, Pilone V, Buonomo T, Liguori R, Castanò I, Buono P et al (2010) miR-519d overexpression is associated with human obesity. Obesity (Silver Spring) 18(11):2170–2176
39. Kloting N, Berthold S, Kovacs P, Schön MR, Fasshauer M, Ruschke K, Stumvoll M, Blüher M (2009) MicroRNA expression in human omental and subcutaneous adipose tissue. PLoS ONE 4(3):e4699
40. Heneghan HM, Miller N, McAnena OJ, O'Brien T, Kerin MJ (2011) Differential miRNA expression in omental adipose tissue and in the circulation of obese patients identifies novel metabolic biomarkers. J Clin Endocrinol Metab 96(5):e846 − 50
41. He Z, Yu J, Zhou C, Ren G, Cong P, Mo D, Chen Y, Liu X (2013) MiR-143 is not essential for adipose development as revealed by in vivo antisense targeting. Biotechnol Lett 35(4):499–507
42. Takanabe R, Ono K, Abe Y, Takaya T, Horie T, Wada H, Kita T, Satoh N, Shimatsu A, Hasegawa K (2008) Up-regulated expression of microRNA-143 in association with obesity in adipose tissue of mice fed high-fat diet. Biochem Biophys Res Commun 376(4):728–732
43. Jordan SD, Krüger M, Willmes DM, Redemann N, Wunderlich FT, Brönneke HS, Merkwirth C et al (2011) Obesity-induced overexpression of miRNA-143 inhibits insulin-stimulated AKT activation and impairs glucose metabolism. Nat Cell Biol 13(4):434–446
44. Kim YJ, Hwang SJ, Bae YC, Jung JS (2009) MiR-21 regulates adipogenic differentiation through the modulation of TGF-beta signaling in mesenchymal stem cells derived from human adipose tissue. Stem Cells 27(12):3093–3102
45. Ling HY, Wen GB, Feng SD, Tuo QH, Ou HS, Yao CH, Zhu BY et al (2011) MicroRNA-375 promotes 3T3-L1 adipocyte differentiation through modulation of extracellular signal-regulated kinase signalling. Clin Exp Pharmacol Physiol 38(4):239–246
46. Kennell JA, Gerin I, MacDougald OA, Cadigan KM (2008) The microRNA miR-8 is a conserved negative regulator of Wnt signaling. Proc Natl Acad Sci U S A 105(40):15417–15422
47. Sun T, Fu M, Bookout AL, Kliewer SA, Mangelsdorf DJ (2009) MicroRNA let-7 regulates 3T3-L1 adipogenesis. Mol Endocrinol 23(6):925–931
48. Yang Z, Bian C, Zhou H, Huang S, Wang S, Liao L, Zhao RC (2011) MicroRNA hsa-miR-138 inhibits adipogenic differentiation of human adipose tissue-derived mesenchymal stem cells through adenovirus EID-1. Stem Cells Dev 20(2):259–267
49. Bussing I, Slack FJ, Grosshans H (2008) let-7 microRNAs in development, stem cells and cancer. Trends Mol Med 14(9):400–409
50. Anand A, Chada K (2000) In vivo modulation of Hmgic reduces obesity. Nat Genet 24(4):377–380
51. Frost RJ, Olson EN (2011) Control of glucose homeostasis and insulin sensitivity by the Let-7 family of microRNAs. Proc Natl Acad Sci U S A 108(52):21075–21080
52. Kinoshita M, Ono K, Horie T, Nagao K, Nishi H, Kuwabara Y, Takanabe-Mori R, Hasegawa K, Kita T, Kimura T (2010) Regulation of adipocyte differentiation by activation of serotonin (5-HT) receptors 5-HT2AR and 5-HT2CR and involvement of microRNA-448-mediated repression of KLF5. Mol Endocrinol 24(10):1978–1987
53. Andersen DC, Jensen CH, Schneider M, Nossent AY, Eskildsen T, Hansen JL, Teisner B, Sheikh SP (2010) MicroRNA-15a fine-tunes the level of Delta-like 1 homolog (DLK1) in proliferating 3T3-L1 preadipocytes. Exp Cell Res 316(10):1681–1691
54. Wellen KE, Hotamisligil GS (2003) Obesity-induced inflammatory changes in adipose tissue. J Clin Invest 112(12):1785–1788
55. Cawthorn WP, Sethi JK (2008) TNF-alpha and adipocyte biology. FEBS Lett 582(1):117–131
56. Wang YC, Li Y, Wang XY, Zhang D, Zhang H, Wu Q, He YQ et al (2013) Circulating miR-130b mediates metabolic crosstalk between fat and muscle in overweight/obesity. Diabetologia 56(10):2275–2285

57. Parra P, Serra F, Palou A (2010) Expression of adipose microRNAs is sensitive to dietary conjugated linoleic acid treatment in mice. PLoS ONE 5(9):e13005
58. Karolina DS, Armugam A, Tavintharan S, Wong MT, Lim SC, Sum CF, Jeyaseelan K (2011) MicroRNA 144 impaired insulin signaling by inhibiting the expression of insulin receptor substrate 1 in type 2 diabetes mellitus. PLoS ONE 6(8):e22839
59. Plaisance V, Abderrahmani A, Perret-Menoud V, Jacquemin P, Lemaigre F, Regazzi R (2006) MicroRNA-9 controls the expression of Granuphilin/Slp4 and the secretory response of insulin-producing cells. J Biol Chem 281(37):26932–26942
60. Poy MN, Eliasson L, Krutzfeldt J, Kuwajima S, Ma X, Macdonald PE, Pfeffer S et al (2004) A pancreatic islet-specific microRNA regulates insulin secretion. Nature 432(7014):226–230
61. Tang X, Muniappan L, Tang G, Ozcan S (2009) Identification of glucose-regulated miRNAs from pancreatic {beta} cells reveals a role for miR-30d in insulin transcription. RNA 15(2):287–293
62. Zhu H, Shyh-Chang N, Segrè AV, Shinoda G, Shah SP, Einhorn WS, Takeuchi A et al (2011) The Lin28/let-7 axis regulates glucose metabolism. Cell 147(1):81–94
63. Pandey AK, Verma G, Vig S, Srivastava S, Srivastava AK, Datta M (2011) miR-29a levels are elevated in the db/db mice liver and its overexpression leads to attenuation of insulin action on PEPCK gene expression in HepG2 cells. Mol Cell Endocrinol 332(1-2):125–133
64. Trajkovski M, Hausser J, Soutschek J, Bhat B, Akin A, Zavolan M, Heim MH, Stoffel M (2011) MicroRNAs 103 and 107 regulate insulin sensitivity. Nature 474(7353):649–653
65. Esguerra JL, Bolmeson C, Cilio CM, Eliasson L (2011) Differential glucose-regulation of microRNAs in pancreatic islets of non-obese type 2 diabetes model Goto-Kakizaki rat. PLoS ONE 6(4):e18613
66. Lovis P, Roggli E, Laybutt DR, Gattesco S, Yang JY, Widmann C, Abderrahmani A, Regazzi R (2008) Alterations in microRNA expression contribute to fatty acid-induced pancreatic beta-cell dysfunction. Diabetes 57(10):2728–2736
67. Zhu Y, You W, Wang H, Li Y, Qiao N, Shi Y, Zhang C, Bleich D, Han X (2013) MicroRNA-24/MODY gene regulatory pathway mediates pancreatic beta-cell dysfunction. Diabetes 62(9):3194–3206
68. Poy MN, Hausser J, Trajkovski M, Braun M, Collins S, Rorsman P, Zavolan M, Stoffel M (2009) miR-375 maintains normal pancreatic alpha- and beta-cell mass. Proc Natl Acad Sci USA 106(14):5813–5818
69. Sun LL, Jiang BG, Li WT, Zou JJ, Shi YQ, Liu ZM (2011) MicroRNA-15a positively regulates insulin synthesis by inhibiting uncoupling protein-2 expression. Diabetes Res Clin Pract 91(1):94–100
70. Fred RG, Bang-Berthelsen CH, Mandrup-Poulsen T, Grunnet LG, Welsh N (2010) High glucose suppresses human islet insulin biosynthesis by inducing miR-133a leading to decreased polypyrimidine tract binding protein-expression. PLoS ONE 5(5):e10843
71. Lovis P, Gattesco S, Regazzi R (2008) Regulation of the expression of components of the exocytotic machinery of insulin-secreting cells by microRNAs. Biol Chem 389(3):305–312
72. Ramachandran D, Roy U, Garg S, Ghosh S, Pathak S, Kolthur-Seetharam U (2011) Sirt1 and mir-9 expression is regulated during glucose-stimulated insulin secretion in pancreatic beta-islets. FEBS J 278(7):1167–1174
73. Baroukh N, Ravier MA, Loder MK, Hill EV, Bounacer A, Scharfmann R, Rutter GA, Van Obberghen E (2007) MicroRNA-124a regulates Foxa2 expression and intracellular signaling in pancreatic beta-cell lines. J Biol Chem 282(27):19575–19588
74. Aspinwall CA, Qian WJ, Roper MG, Kulkarni RN, Kahn CR, Kennedy RT (2000) Roles of insulin receptor substrate-1, phosphatidylinositol 3-kinase, and release of intracellular Ca2+ stores in insulin-stimulated insulin secretion in beta -cells. J Biol Chem 275(29):22331–22338
75. El Ouaamari A, Baroukh N, Martens GA, Lebrun P, Pipeleers D, van Obberghen E (2008) miR-375 targets 3'-phosphoinositide-dependent protein kinase-1 and regulates glucose-induced biological responses in pancreatic beta-cells. Diabetes 57(10):2708–2717

76. Michael MD, Kulkarni RN, Postic C, Previs SF, Shulman GI, Magnuson MA, Kahn CR (2000) Loss of insulin signaling in hepatocytes leads to severe insulin resistance and progressive hepatic dysfunction. Mol Cell 6(1):87–97
77. Burks DJ MF (2001) White, IRS proteins and beta-cell function. Diabetes 50(Suppl 1):S140–S145
78. Shepherd PR, Withers DJ, Siddle K (1998) Phosphoinositide 3-kinase: the key switch mechanism in insulin signalling. Biochem J 333(Pt 3):471–490
79. Herrera BM, Lockstone HE, Taylor JM, Ria M, Barrett A, Collins S, Kaisaki P et al (2010) Global microRNA expression profiles in insulin target tissues in a spontaneous rat model of type 2 diabetes. Diabetologia 53(6):1099–1109
80. Herrera BM, Lockstone HE, Taylor JM, Wills QF, Kaisaki PJ, Barrett A, Camps C et al (2009) MicroRNA-125a is over-expressed in insulin target tissues in a spontaneous rat model of type 2 Diabetes. BMC Med Genomics 2:54
81. He A, Zhu L, Gupta N, Chang Y, Fang F (2007) Overexpression of micro ribonucleic acid 29, highly up-regulated in diabetic rats, leads to insulin resistance in 3T3-L1 adipocytes. Mol Endocrinol 21(11):2785–2794
82. Kornfeld JW, Baitzel C, Könner AC, Nicholls HT, Vogt MC, Herrmanns K, Scheja L et al (2013) Obesity-induced overexpression of miR-802 impairs glucose metabolism through silencing of Hnf1b. Nature 494(7435):111–115
83. Ryu HS, Park SY, Ma D, Zhang J, Lee W (2011) The induction of microRNA targeting IRS-1 is involved in the development of insulin resistance under conditions of mitochondrial dysfunction in hepatocytes. PLoS ONE 6(3):e17343
84. Davalos A, Goedeke L, Smibert P, Ramírez CM, Warrier NP, Andreo U, Cirera-Salinas D et al (2011) miR-33a/b contribute to the regulation of fatty acid metabolism and insulin signaling. Proc Natl Acad Sci U S A 108(22):9232–9237
85. Roush S, Slack FJ (2008) The let-7 family of microRNAs. Trends Cell Biol 18(10):505–516
86. Xu J, Wong C (2008) A computational screen for mouse signaling pathways targeted by microRNA clusters. RNA 14(7):1276–1283
87. Hunter MP, Ismail N, Zhang X, Aguda BD, Lee EJ, Yu L, Xiao T et al (2008) Detection of microRNA expression in human peripheral blood microvesicles. PLoS ONE 3(11):e3694
88. Turchinovich A, Weiz L, Langheinz A, Burwinkel B (2011) Characterization of extracellular circulating microRNA. Nucleic Acids Res 39(16):7223–7233
89. Kosaka N, Iguchi H, Yoshioka Y, Takeshita F, Matsuki Y, Ochiya T (2010) Secretory mechanisms and intercellular transfer of microRNAs in living cells. J Biol Chem 285(23):17442–17452
90. Vickers KC, Palmisano BT, Shoucri BM, Shamburek RD, Remaley AT (2011) MicroRNAs are transported in plasma and delivered to recipient cells by high-density lipoproteins. Nat Cell Biol 13(4):423–433
91. Laterza OF, Lim L, Garrett-Engele PW, Vlasakova K, Muniappa N, Tanaka WK, Johnson JM et al (2009) Plasma MicroRNAs as sensitive and specific biomarkers of tissue injury. Clin Chem 55(11):1977–1983
92. Fichtlscherer S, De Rosa S, Fox H, Schwietz T, Fischer A, Liebetrau C, Weber M et al (2010) Circulating microRNAs in patients with coronary artery disease. Circ Res 107(5):677–684
93. Chen X, Ba Y, Ma L, Cai X, Yin Y, Wang K, Guo J et al (2008) Characterization of microRNAs in serum: a novel class of biomarkers for diagnosis of cancer and other diseases. Cell Res 18(10):997–1006
94. Mitchell PS, Parkin RK, Kroh EM, Fritz BR, Wyman SK, Pogosova-Agadjanyan EL, Peterson A et al (2008) Circulating microRNAs as stable blood-based markers for cancer detection. Proc Natl Acad Sci U S A 105(30):10513–10518
95. Cermelli S, Ruggieri A, Marrero JA, Ioannou GN, Beretta L (2011) Circulating microRNAs in patients with chronic hepatitis C and non-alcoholic fatty liver disease. PLoS ONE 6(8):e23937

96. Weiland M, Gao XH, Zhou L, Mi QS (2012) Small RNAs have a large impact: circulating microRNAs as biomarkers for human diseases. RNA Biol 9(6):850–859
97. Zampetaki A, Kiechl S, Drozdov I, Willeit P, Mayr U, Prokopi M, Mayr A et al (2010) Plasma microRNA profiling reveals loss of endothelial miR-126 and other microRNAs in type 2 diabetes. Circ Res 107(6):810–817
98. Krutzfeldt J, Rajewsky N, Braich R, Rajeev KG, Tuschl T, Manoharan M, Stoffel M (2005) Silencing of microRNAs in vivo with 'antagomirs'. Nature 438(7068):685–689 (ADDIN EN.REFLIST)

Chapter 10
Obesity Study: Animal Models

Goro Katsuura, Namiko Kawamura, Miyuki Nishida, Haruka Amitani, Akihiro Asakawa and Akio Inui

Introduction

Body weight and adiposity are maintained at near-constant levels through matching food intake to energy expenditure over long periods. Obesity is defined as a high body mass index with a large amount of adiposity resulting from chronic excess of energy intake over energy expenditure. Recently, the number of obese people increased in the world, and obesity is becoming a world-wide problem because obesity is a risk factor in increased induction of type II diabetes, cancer, hyperlipidemia and hypertension, so-called metabolic syndrome [1]. Moreover, clinical studies have revealed that obesity is comorbid with several forms of mental disorder [2–4]. Epidemiological studies have demonstrated that the incidence of depression and cognitive impairment is higher in obese subjects than normal body weight subjects [5, 6]. There is the possibility that mental disorder acts as a trigger of the development of obesity. Also, such mental disorder might cause further progression of obesity. After findings on the hypothalamus as the center of energy regulation in 1940's, the central nervous system came to the forefront of attention in the pathophysiology of obesity. Genetic and environmental factors play a role in the development of obesity, and diet is one of the main environmental factors that contribute to this disease [1, 7]. Human studies have shown that increased fat intake is associated with body weight gain, resulting in obesity and other related metabolic diseases. It is therefore important to understand the basics of the controls of food intake and how they might relate to obesity. Animal rodent models are useful tools for studying obesity, and are important in the development of treatments for obesity. This chapter shows representative animal obese models that are used to analyze pathophysiological properties of obesity.

A. Inui (✉) · G. Katsuura · N. Kawamura · M. Nishida · H. Amitani · A. Asakawa
Department of Psychosomatic Internal Medicine, Kagoshima University Graduate School of Medical and Dental Sciences, 8-35-1 Sakuragaoka, Kagoshima 890-8520, Japan
e-mail: inui@m.kufm.kagoshima-u.ac.jp

C. Nóbrega, R. Rodriguez-López (eds.), *Molecular Mechanisms Underpinning the Development of Obesity*, DOI 10.1007/978-3-319-12766-8_10

Energy Regulation

Energy homeostasis including food intake and energy consumption has been demonstrated to be predominantly regulated by orexigenic and anorexigenic systems in the hypothalamus (Fig. 10.1) [8–10]. The blood-born signals, such as leptin, insulin and ghrelin, mainly interact with receptors in the hypothalamus to regulate food intake and energy expenditure [9, 10]. In the arcuate nucleus (ARC), orexigenic neuropeptides, such as neuropeptide Y (NPY) and agouti-related protein (AgRP) increase food intake and decrease energy expenditure. On the other hand, anorexigenic neuropeptides, α-melanocyte-stimulating hormone (α-MSH) and cocaine- and amphetamine-regulated transcript (CART) decrease food intake and increase energy expenditure. Activities of these neuropeptides in the ARC are regulated by leptin, insulin and ghrelin. These neuropeptide-containing neurons project to lateral hypothalamic area (LHA), ventromedial hypothalamus (VMH), dorsomedial hypothalamus (DMH) and to paraventricular nucleus (PVN). Moreover, the LHA, orexigenic neuropeptides, such as orexin (ORX) and melanin-concentrating hormone (MCH), project to mesolimbic dopamine system, the so-called reward system. Oxytocin (OXT) neurons project to the brain stem and modulate activity of sympathetic and parasympathetic nervous systems. Recently, several lines of evidence have indicated that energy regulation is also modulated by extra-hypothalamic brain areas originally related to regulation of emotion and cognition, such as the nucleus

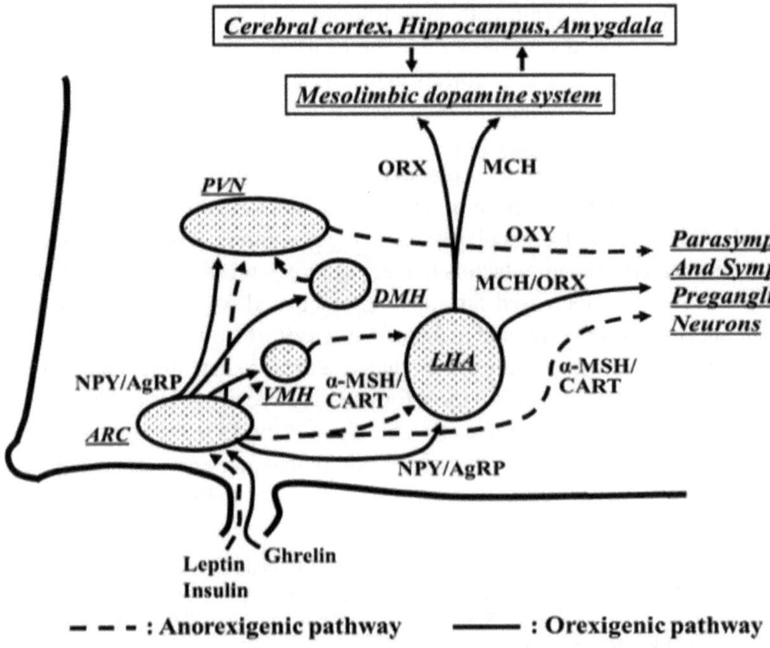

Fig. 10.1 Energy regulation in the central nervous system

accumbens, amygdala, hippocampus and cerebral cortex [9]. These findings suggest that maintaining energy homeostasis and regulating emotion and cognition share common brain regions, as well as bidirectional interaction between energy regulation and emotional/cognitive functions.

Animal Models

Genetic Obese Models

Lethal Yellow Mutant Mouse (Ay)

An agouti mouse was first described more than a century ago. Agouti gene is the first obesity gene characterized at the molecular level [11]. The agouti protein transiently expressed in the follicular melanocyte stimulates production of the red/yellow pigment and inhibits production of the black/brown pigment [12, 13]. The red/yellow pigment production is especially regulated by antagonism of the binding of α-MSH to melanocortin 1 receptor (MC1R) [12]. The lethal yellow mutant mouse (Ay) is one of five dominant agouti mutations and has been found to be an excellent mouse model of obesity [11]. The Ay mutation is characterized by the deletion of 120–170 kb genomic DNA that places it under the control of a ubiquitously expressed promoter, resulting in ubiquitous agouti expression due to loss of the tissue-specific control promoter element [14–16]. The Ay mouse has a complex phenotype including yellow coat color, mature-onset obesity, type II diabetes, hyperleptinemia, increased linear growth, higher tumor susceptibility, and infertility. Transgenic mice expressing ubiquitous agouti exhibited yellow coat color, obesity, hyperinsulinemia, and hyperglycemia similar to Ay mouse [15], revealing the molecular mechanism of agouti in mouse phenotype. However, agouti expression in skin has little effect on the obesity phenotype, because transgenic mice overexpressing agouti in skin do not become obese [17]. Mice with adipose tissue-specific agouti overexpression exhibit an overgrowth of adipose tissue without alteration of food intake, suggesting that increased fat in this model is due to changes in energy metabolism [18]. The adipose tissue agouti overexpression model could be relevant to human obesity because agouti gene expression is found in human adipose tissue [19, 20] and is increased in the adipose tissue of type 2 diabetic subjects [21]. The obesogenic role of agouti is considered to be tissue dependent [22]. The molecular categorization of agouti was responsible for elucidation of the melanocortin system's involvement in energy regulation, due to its mimicry of AgRP activity in the hypothalamus. The melanocortin receptor family comprises five G-protein–coupled proteins, MC1R to MC5R, which demonstrate tissue-specific patterns of expression. Especially, MC4R in the hypothalamus plays a key role in the regulation of feeding and metabolism and is normally antagonized by AgRP [23]. Novel expression of agouti in the brain results in the chronic antagonism of MC4R, disrupting its function.

Ob/ob Mouse (Lepob/Lepob mouse)

In 1950, Ingalls et al. described a spontaneous mutation in a V mice colony, which provoked severe obesity. This mutation, named ob [24], was subsequently introduced into C57BL/6J to obtain ob/ob mice, which are obese and exhibit hyperinsulinemia, insulin resistance, hyperglycemia, infertility, hypothyroidism, hypercorticoidism, low sympathetic activity, impaired thermoregulation and lowered physical activity [25, 26]. In 1966, Hummel et al. described a spontaneous mutation in a C57BL/Ks mice colony that provoked moderate hyperphagia and obesity, accompanied by severe fasting hyperglycemia increasing with age and a marked increase in plasma insulin concentration [27, 28]. Parabiosis studies with ob/ob mice and normal mice indicated that the mutation causes a deficiency in a circulating lipostatic factor [26]. In 1994, Zhang et al. [29] identified the product of the gene mutation in ob/ob mice that was responsible for their obesity and was also the presumed circulating factor. This protein, leptin, has the structure of a long chain helical cytokine [30] and is expressed in adipose tissue in proportion to adipocyte size [31, 32]. Leptin is an adipocyte-derived blood-borne satiety factor that is involved in the regulation of food intake and body weight acting in the hypothalamus. Ob/ob mice exhibit a mutation located on the +105 codon, replaces an arginine by a stop codon (CGA→TGA) [29], resulting in a deficiency of biologically functional leptin. It was reported that leptin recovers the decreased metabolic rate, body temperature, and locomotor activity in ob/ob mice [33]. The discovery of leptin has opened up a whole new field of studies on regulation of food intake, energy expenditure and obesity. They have been used in many studies of the effects of antiobesity and antidiabetogenic drugs.

Db/db Mouse (Leprdb/Lepdb Mouse)

The db (stands for "diabetes") mutation is an autosomal recessive trait. The obese mouse was identified initially in 1966 by researchers in the Jackson Laboratory [34]. In 1995, the mouse leptin receptor gene was cloned [35] and the db gene encodes for a G-to-T point mutation of the leptin receptor, leading to abnormal splicing and defective signaling of the adipocyte-derived hormone leptin [36, 37]. This mutation affects the alternative splicing of the ob-Rb subtype of leptin receptors, the only subtype with a functional C-terminal intracellular domain to transmit the signal. Leptin can bind to the receptors but the signal transduction system is not functional. Lack of leptin signaling in the hypothalamus leads to persistent hyperphagia, hyperlipidemia, insulin resistance and diabetes [37, 38]. Following the onset of overt diabetes, db/db mice develop progressive kidney disease, which has similarities to human diabetic nephropathy. Therefore, db/db mice are now regularly used to examine the mechanisms of renal injury, which result from type 2 diabetes.

SHROB (Spontaneously Hypertensive-Obese) rat

SHROB rats were first described in 1973 [39]. They result from crossing a spontaneously hypertensive female rat and a Sprague–Dawley male rat, which leads to a spontaneous mutation, first designated f for fatty [40, 41] and then fak [42]. This mutation is recessive and is located on the +763 codon of the leptin receptor gene in a sequence coding for the extracellular domain of the leptin receptor and common to all subtypes. This mutation is non-sense, replacing a tyrosine with a stop codon (TAT→TAA), and does not affect the leptin receptor mRNA. However, the protein is not expressed [43–45]. They are obese because of the accumulation of fat, in particular at the sub-cutaneous, retroperitoneal and mesenteric level. These animals exhibit hyperlipidemia, hyperglycemia and insulin resistance. Arterial pressure in SHROB rats increases with age up to 180 mmHg for systolic blood pressure. The life span of male SHROB is quite short (10–11 months in average) and the principal causes of death are kidney, urinary tract or vascular pathologies [40, 41].

Zucker Fatty Rat and Zucker Diabetic Fatty Rat

In 1961, Zucker et al. observed an obese phenotype due to a spontaneous mutation in their 13 M rat colony [46, 47]. The mutation, named fatty or fa, is an autosomal recessive mutation in the fatty (fa) gene and affects the leptin receptor gene, but is different from the fak mutation of SHROB rats [42]. Located on the +269 codon, it replaces a glutamine with a proline, leading to production of a truncated protein [43, 44, 48, 49]. This mutation affects the extracellular part of the leptin receptor. In experiments using cells expressing wild-type or mutated leptin receptors, mutated receptors have shown weaker affinity for leptin, and altered signal transduction [50, 51]. Zucker fatty rats required much higher doses of leptin than normal rats to produce a similar effect [52]. These rats are characterized by hyperphagia and early-onset obesity corresponding to an accumulation of subcutaneous and retroperitoneal fat. Zucker fatty rats exhibit hyperlipidemia with age. They are slightly insulin-resistant but do not exhibit fasting hyperglycemia. Nevertheless, obesity in Zucker fatty rats is not as marked as in SHROB rats. Zucker fatty rats can present moderate arterial hypertension, but only when old [46, 47, 53]. In addition, Zucker diabetic fatty (ZDF) rats were derived from substrain of obese Zucker fatty rats and display early deregulation of glucose metabolism [54, 55]. ZDF rats constitute a model of metabolic syndrome with type 2 diabetes.

OLETF Rat (Otsuka Long-Evans Tokushima Fatty Rat)

OLETF rats were developed by the selection of spontaneously type 2 diabetic rats from the outbreeding of Long Evans rats in a closed colony of Charles River at the Tokushima Research Institute of Otsuka Pharmaceutical in Japan. [56]. OLETF rats are hyperphagic beginning several weeks after birth, with increasing body weight eventually progressing to frank obesity [56] These rats exhibit obesity, hyperglycemia, hypertriglyceridemia, hyperinsulinemia and chronic diabetes mellitus [56, 57]. In 1995, Otsuki et al. demonstrated that pancreatic acini of OLETF rats are insensitive to the amylase actions of cholecystokinin (CCK) due to low expression or absence of expression of CCK-1 receptor mRNA [58, 59]. Other studies showed that this lack of expression of CCK-1 receptors is due to a deletion including the promoter region and the first and second exons [60]. Because CCK plays an important role in satiation, OLETF rats are a valuable animal model to study deregulated control of eating and obesity.

Diet-Induced Model

High-Fat Diet-Induced Obese Model

The development of obesity and metabolic syndrome in human is mostly linked with increased caloric intake and lack of physical activity, in addition to genetic predisposition. It would be of interest to study the pathogenesis of metabolic syndrome induced by greater food intake in general. A high-fat diet is often used in obesity research as a non-leptin-deficient model. In animal models, although there is no consensus about the definition of a low- or high-fat diet, and a wide variety of diets are used in animal experiments, in general, standard diets contain less than 10 % of calories from fat whereas high-fat diets or very-high-fat diets contain 30–50 %, and more than 50 % of calories from fat, respectively. From a nutritional perspective, diets with 60 kcal% fat are commonly used to induce obesity in rodents since animals tend to gain more weight more quickly thereby allowing researchers to screen their compounds after a shorter period of time [61, 62]. Many high-fat diets used in laboratory animal research contain more saturated fat such as lard, beef tallow, or coconut oil and are quite capable of inducing metabolic disorders such as obesity, insulin resistance or glucose intolerance in susceptible strains [63–65]. Addition of sucrose aggravates the metabolic consequences of enriched diets and induces severe dyslipidemia [66, 67]. Moreover, high-fat diets rapidly and specifically reduce the central actions of insulin and leptin, most likely due to a post-receptor effect [68–72]. This effect is rapid, occurring after a few days of HF exposure.

An appropriate diet in an appropriate strain is necessary to induce the development of a specific metabolic problem because there are mouse strain-specific differences in responses to the high-fat diet [73]. Among the various strains, C57BL/6J mice are the most widely used for high-fat diet -induced obesity because they exhibit abnormalities similar to human metabolic syndrome when fed with high-fat diet [74]. Interestingly, within the C57 mouse strain, there are significant differences among substrains in response to the high-fat diet. For instance, whereas C57BL/6J mice exhibit high-fat diet -induced obesity, hyperinsulinemia, and insulin resistance that closely parallel the progression of human disease, C57BL/KsJ mice display a weak phenotype [73]. For example, murine strains C57BL/6 and AKR both develop obesity on a high-fat diet, but (for the same weight gain) C57BL/6 mice also exhibit fasted hyperglycemia, and glucose intolerance due to a decreased insulin concentration, whereas AKR mice also exhibit hyperinsulinemia but with normal fasted glucose [75]. Other strains, such as A/J mice, are resistant to obesity induced by a high-fat diet [66]. Rat models including Sprague-Dawley and Wistar rats are popular strains to study obesity as they readily gain weight on high-fat diets. Sprague Dawley or Long Evans rats are also used for non-mouse rodent models of high-fat diet -induced obesity [76].

These reports indicated that the high-fat diet-induced obesity animals could be a good model for the experimental therapy and the translational research aiming to discover a novel therapeutic strategy for obesity epidemic.

Cafeteria Diet-Induced Obese Model

Another experimental rodent diet model that exists, which more accurately reflects the variety of highly palatable, energy dense foods that are prevalent in Western society and associated with the current obesity pandemic: the "cafeteria diet". The cafeteria diet foods include cookies, cereals, cheese, processed meats, crackers, etc. Rats become obese when offered a varied and palatable diet that mimics the so-called Western diet of humans (cafeteria diet) [77–79]. Cafeteria diet-induced obesity mainly results from hyperphagia with increased average meal size as well as increased meal frequency that is partly compensated by increased energy expenditure, in particular diet-induced thermogenesis due to sympathetic activation of brown fat. This contrasts with overeating of palatable diets with no choice of foods, which mainly influences meal size.

Genetically Modified Obese Model

As the central and peripheral pathways involved in food intake and energy expenditure are elucidated, a wide range of genes affecting these processes is being identified. Accordingly, many genetically modified obese mice have been created with genes either overexpressed or deleted. In particular, genes of the melanocortin system have been examined using this approach because melanocortin system is a major player in regulation of energy homeostasis (Fig. 10.1).

POMC (Proopiomelanocortin)-Knockout Mouse

POMC-expressing neurons in the arcuate nucleus of the hypothalamus are direct targets of leptin. POMC is the precursor of several biologically active peptides including α-MSH. In the brain, α-MSH is a potent anorexigenic neuropeptide that reduces food intake and increases energy expenditure by activating MC3R and MC4R in the paraventricular nucleus of the hypothalamus and elsewhere. Mice lacking POMC overeat and develop marked obesity that is exaggerated on a high-fat diet [80, 81]. Heterozygous mutants develop an intermediate phenotype, implying that a functional POMC gene is necessary to maintain normal energy homeostasis. Although treatment with leptin is ineffective in POMC-knockout mice, the obesity in POMC-knockout mice can be markedly reduced when these mice are treated with α-MSH or other agonists for the MC4 receptor, such as MT II. POMC deficiency has also been reported in rare cases of human obesity [82].

MC4R and MC3R-Knockout Mouse

α-MSH and AgRP influence energy homeostasis via melanocortin receptors. The MC4 receptor subtype among melanocortin receptors in particular is involved in the control of food intake. These mice exhibit obesity, hyperphagia, hyperglycemia, hyperinsulinemia [83, 84]. Knockout of the MC4R gene in mice is observed to result in early-onset obesity, non–insulin-dependent diabetes and other obesity associated syndromes. As agouti peptide is an antagonist of MC4-R, these symptoms are a parallel to the yellow agouti mouse syndrome, indicating that agouti expression in the hypothalamus inhibits MC4R function, leading to obesity. In contrast to many other obesity models, MC4R-knockout mice do not have elevated circulating corticosterone levels. MC4-knockout mice do not respond to leptin, AgRP or α-MSH. Similar mutations of the MC4 receptor are often stated to be the most frequent genetic cause of obesity in humans. An MC4R knockout rat has recently been described [85]. It has many characteristics in common with MC4R-knockout mice (such as increased body weight, food intake and body length, and lower spontaneous activity). Targeted deletion of the MC3R gene also results in a late onset obesity phenotype, but regulation of appetite and metabolism appear to be intact [86].

Hypothalamic Lesion-Induced Obese Model

In the earliest obesity study with rodents, obese models in rats were induced by surgical or chemical lesions of the VMH as a satiety center and the ARC of the hypothalamus that results in hyperphagia, increased body weight and adiposity.

VMH Lesion-Induced Obese Model

Rats with VMH lesions often begin eating voraciously even before fully recovering from the effects of anesthesia [87, 88]. The overeating and obesity have traditionally been divided into two stages [88]: a dynamic phase of marked hyperphagia and rapid weight gain followed by declining food intake as body weight is maintained during the static phase of obesity. While the precise mechanisms underlying VMH-lesion induced obesity are still unclear, it was detected a change in the tone of the sympathetic (decrease) and parasympathetic (increase) nervous systems that is associated with reduced energy expenditure contributes to the syndrome [89]. VMH lesion induces hyperphagia, resulting in hyperglycemia, hyperinsulinemia, insulin resistance, and reduction in physical activity, hyperlipidemia and obesity. Hyperphagia is probably due to the destruction of POMC neurons from the ARC and possibly of neurons producing brain-derived neurotrophic factor as anorexigenic factor in the VMH [90–92]. VMH lesion-induced obesity has been found in a wide variety of species including mice, ground squirrels, rabbits, cats, dogs, pigs, goats, chickens, sparrows, monkeys, and humans [90].

ARC Lesion-Induced Obese Model

The ARC is considered one of the most important hypothalamic regions involved in the regulation of energy homoeostasis. It is difficult to perform a selective surgical lesion of the ARC due to its anatomical shape and location. The repeated administration of monosodium glutamate (MSG), which is a neuroexcitatory amino acid that is harmful to the central nervous system (CNS), to neonatal rats within the first 10 postnatal days has been used to induce the relatively distinct lesion of ARC neurons projecting to the VMH and PVN. All features of metabolic syndrome, such as massive adipose tissue accumulation, insulin resistance, glucose intolerance, hypogonadism and dyslipidaemia are induced by neonatal MSG treatment [93–96]. It is important to note that systemic MSG treatment also lesions neurons in the circumventricular organs due to their open blood-brain barrier. MSG lesions are therefore not restricted to the ARC and interpretation of results must take this into account. ARC neurons can also be destroyed by local administration of goldthioglucose, resulting in a similar obesity phenotype [97–99].

Conclusion

There has been a growing body of literature using rodents as models of human obesity, even though there are many confounding factors including species, strain, age of the animals, type of diet, level of fat, and type of control diet. The development of the human obesity is multifactorial; it depends not only on particular environmental conditions (increased caloric intake, activity), but also on genetic predisposition, moreover has a strong component of individual dependence. This aspect is probably a limit of the use of animal models to investigate human obesity and metabolic syndrome. Therefore, intensive and continuous efforts should be made to establish novel obesity-associated animal models that mimic human health problems, as suitable animal models are fundamental to testing novel therapeutic strategies against obesity.

References

1. Kopelman PG (2000) Obesity as a medical problem. Nature 404:635–643
2. Simon GE, Von Korff M, Saunders K et al (2006) Association between obesity and psychiatric disorders in the US adult population. Arch Gen Psychiatry 63:824–830
3. Malnick SD, Knobler H (2006) The medical complications of obesity. QJM 99:565–579
4. Marcus MD, Wildes JE (2009) Obesity: is it a mental disorder? Int J Eat Disord 42:739–753
5. Elias MF, Elias PK, Sullivan LM et al (2003) Lower cognitive function in the presence of obesity and hypertension: the Framingham heart study. Int J Obes 27:260–268
6. Whitmer RA, Gunderson EP, Barrett-Connor E et al (2005) Obesity in middle age and future risk of dementia: a 27 year longitudinal population based study. BMJ 330:1360–1364
7. Barsh GS, Farooqi IS, O'Rahilly S (2000) Genetics of body-weight regulation. Nature 404:644–651
8. Schwartz MW, Woods SC, Porte D Jr et al (2000) Central nervous system control of food intake. Nature 404:661–671
9. Morton1 GJ, Cummings DE, Baskin DG et al (2006) Central nervous system control of food intake and body weight. Nature 443:289–295
10. Joel K, Elmquist JK, Elias CF, Saper CB (1999) From lesions to leptin: hypothalamic control of food intake and body weight. Neuron 22:221–232
11. Bultman SJ, Michaud EJ, Woychik RP (1992) Molecular characterization of the mouse agouti locus. Cell 71:1195–1204
12. Lu D, Willard D, Patel IR et al (1994) Agouti protein is an antagonist of the melanocyte-stimulating-hormone receptor. Nature 371:779–802
13. Millar SE, Miller MW, Stevens ME et al (1995) Expression and transgenic studies of the mouse agouti gene provide insight into mechanisms by which mammalian coat color patterns are generated. Development 121:3223–3232
14. Michaud EJ, Bultman SJ, Stubbs LJ et al (1993) The embryonic lethality of homozygous lethal yellow mice (Ay/Ay) is associated with the disruption of a novel RNA-binding protein. Gene Dev 7:1203–1213
15. Duhl DM, Stevens ME, Vrieling H et al (1994) Pleiotropic effects of the mouse lethal yellow (Ay) mutation explained by deletion of a maternally expressed gene and the simultaneous production of agouti fusion RNAs. Development 120:1695–1708
16. Michaud EJ, Bultman SJ, Klebig ML et al (1994) A molecular model for the genetic and phenotypic characteristics of the mouse lethal yellow (Ay) mutation. Pro Natl Acad Sci U S A 91:2562–2566
17. Kucera GT, Bortner DM, Rosenberg M (1996) Overexpression of an Agouti cDNA in the skin of transgenic mice recapitulates dominat coat color phenotypes of spontaneous mutants. Dev Biol 173:162–173
18. Mynatt RL, Miltenberger RJ, Klebig ML et al (1997) Combined effects of insulin treatment and adipose tissue-specific agouti expression on the development of obesity. Pro Natl Acad Sci U S A 94:919–922
19. Kwon HY, Bultman SJ, Löffler C et al (1994) Molecular structure and chromosomalmapping of the human homolog of the agouti gene. Proc Natl Acad Sci U S A 91:9760–9764
20. Wilson BD, Ollmann MM, Kang L et al (1995) Structure and function of ASP, the human homolog of the mouse agouti gene. Hum Mol Genet 4:223–230
21. Smith SR, Gawronska-Kozak B, Janderov'a L et al (2003) Agouti expression in human adipose tissue: functional consequences and increased expression in type 2 diabetes. Diabetes 52:2914–2922
22. Kucera GT, Bortner DM, Rosenberg MP (1996) Overexpression of an Agouti cDNA in the skin of transgenic mice recapitulates dominant coat color phenotypes of spontaneous mutants. Dev Biol 173:162–173
23. Cone RD, Lu D, Koppula S et al (1996) The melanocortin receptors: agonists, antagonists, and the hormonal control of pigmentation. Recent Prog Horm Res 51:287–318

24. Ingalls AM, Dickie MM, Snell GD (1950) Obese, a new mutation in the house mouse. J Hered 41:317–318
25. Bray GA, York DA (1979) Hypothalamic and genetic obesity in experimental animals: an autonomic and endocrine hypothesis. Physiol Rev 59:719–809
26. Coleman DL, Hummel KP (1973) The influence of genetic background on the expression of the obese (Ob) gene in the mouse. Diabetologia 9:287–293
27. Hummel KP, Dickie MM, Coleman DL (1966) Diabetes, a new mutation in the mouse. Science 153:1127–1128
28. Coleman DL, Hummel KP (1967) Studies with the mutation, diabetes, in the mouse. Diabetologia 3:238–248
29. Zhang Y, Proenca R, Maffei M et al (1994) Positional cloning of the mouse obese gene and its human homologue. Nature 372:425–432
30. Zhang F, Basinski MB, Beals JM et al (1997) Crystal structure of the obese protein leptin-E100. Nature 387:206–209
31. Hamilton BS, Paglia D, Kwan AY et al (1995) Increased obese mRNA expression in omental fat cells from massively obese humans. Nat Med 1:953–956
32. Harris RB, Ramsay TG, Smith SR et al (1996) Early and late stimulation of ob mRNA expression in meal-fed and overfed rats. J Clin Invest 97:2020–2026
33. Pelleymounter MA, Cullen MJ, Baker MB et al (1995) Effects of the obese gene product on body weight regulation in ob/ob mice. Science 269:540–543
34. Hummel KP, Dickie MM, Coleman DL (1966) Diabetes, a new mutation in the mouse. Science 153:1127–1128
35. Tartaglia LA, Dembski M, Weng X et al (1995) Identification and expression cloning of a leptin receptor, OB-R. Cell 83:1263–1271
36. Chen H, Charlat O, Tartaglia LA et al (1996) Evidence that the diabetes gene encodes the leptin receptor: identification of a mutation in the leptin receptor gene in db/db mice. Cell 84:491–495
37. Lee GH, Proenca R, Montez JM et al (1996) Abnormal splicing of the leptin receptor in diabetic mice. Nature 379:632–635
38. Chen H, Charlat O, Tartaglia LA et al (1996) Evidence that the diabetes gene encodes the leptin receptor: identification of a mutation in the leptin receptor gene in db/db mice. Cell 84:491–495
39. Koletsky S (1973) Obese spontaneously hypertensive rats-a model for study of atherosclerosis. Exp Mol Pathol 19:53–60
40. Koletsky S (1975) Animal model: obese hypertensive rat. Am J Pathol 81(2):463–466
41. Koletsky S (1975) Pathologic findings and laboratory data in a new strain of obese hypertensive rats. Am J Pathol 80:129–142
42. Yen TT, Shaw WN, Yu PL (1977) Genetics of obesity of Zucker rats and Koletsky rats. Heredity 38:373–377
43. Takaya K, Ogawa Y, Hiraoka J et al (1996) Nonsense mutation of leptin receptor in the obese spontaneously hypertensive Koletsky rat. Nat Genet 14:130–131
44. Takaya K, Ogawa Y, Isse N et al (1996) Molecular cloning of rat leptin receptor isoform complementary DNAs—identification of a missense mutation in Zucker fatty (fa/fa) rats. Biochem Biophys Res Commun 225:75–83
45. Ishizuka T, Ernsberger P, Liu S et al (1998) Phenotypic consequences of a nonsense mutation in the leptin receptor gene (fak) in obese spontaneously hypertensive Koletsky rats (SHROB). J Nutr 128:2299–2306
46. Zucker LM (1965) Hereditary obesity in the rat associated with hyperlipemia. Ann N Y Acad Sci 131:447–458
47. Zucker LM, Zucker TF (1961) Fatty, a new mutation in the rat. J Hered 52:275
48. Iida M, Murakami T, Ishida K (1996) Phenotype-linked amino acid alteration in leptin receptor cDNA from Zucker fatty (fa/fa) rat. Biochem Biophys Res Commun 222:19–26

49. Iida M, Murakami T, Ishida K et al (1996b) Substitution at codon 269 (glutamine→proline) of the leptin receptor (OB-R) cDNA is the only mutation found in the Zucker fatty (fa/fa) rat. Biochem Biophys Res Commun 224:597–604
50. White DW, Wang DW, Chua SC et al (1997) Constitutive and impaired signaling of leptin receptors containing the Gln→Pro extracellular domain fatty mutation. Proc Natl Acad Sci U S A 94:10657–10662
51. Yamashita T, Murakami T, Iida M et al (1997) Leptin receptor of Zucker fatty rat performs reduced signal transduction. Diabetes 46:1077–1080
52. Cusin I, Rohner-Jeanrenaud F, Stricker-Krongrad A et al (1996) The weight-reducing effect of an intracerebroventricular bolus injection of leptin in genetically obese fa/fa rats. Reduced sensitivity compared with lean animals. Diabetes 45:1446–1450
53. Tofovic SP, Kusaka H, Kost CK Jr et al (2000) Renal function and structure in diabetic, hypertensive, obese ZDF × SHHF-hybrid rats. Ren Fail 22:387–406
54. Peterson RG, Shaw WN, Neel M-A et al (1990) Zucker diabetic fatty rat as a model for non-insulin-dependent diabetes mellitus. ILAR J 32:16–19
55. Schlenker EH, Tamura T, Gerdes AM (2003) Gender-specific effects of thyroid hormones on cardiopulmonary function in SHHF rats. J Appl Physiol 95:2292–2298
56. Kawano K, Hirashima T, Mori S et al (1992) Spontaneous long-term hyperglycemic rat with diabetic complications: Otsuka long-evans tokushima fatty (OLETF) strain. Diabetes 41:1422–1428
57. Moran TH, Bi S (2006) Hyperphagia and obesity in OLETF rats lacking CCK-1 receptors. Philos Trans R Soc Lond B Biol Sci 361:1211–1218
58. Funakoshi A, Miyasaka K, Jimi A et al (1994) Little or no expression of the cholecystokinin-A receptor gene in the pancreas of diabetic rats (Otsuka long-evans tokushima fatty = OLETF rats). Biochem Biophys Res Commun 199:482–488
59. Otsuki M, Akiyama T, Shirohara H et al (1995) Loss of sensitivity to cholecystokinin stimulation of isolated pancreatic acini from genetically diabetic rats. Am J Physiol 268:E531–E536
60. Takiguchi S, Takata Y, Funakoshi A et al (1997) Disrupted cholecystokinin type-A receptor (CCKAR) gene in OLETF rats. Gene 197:169–175
61. Ghibaudi L, Cook J, Farley C et al (2002) Fat intake affects adiposity, comorbidity factors, and energy metabolism of sprague-dawley rats. Obes Res 10:956–963
62. Johnston SL, Souter DM, Tolkamp BJ et al (2007) Intake compensates for resting metabolic rate variation in female C57BL/6J mice fed high-fat diets. Obesity 15:600–606
63. Ikemoto S, Takahashi M, Tsunoda N et al (1996) High-fat diet-induced hyperglycemia and obesity inmice: differential effects of dietary oils. Metabolism 45:1539–1546
64. Wang H, Storlien LH, Huang XF (2002) Effects of dietary fat types on body fatness, leptin, and ARC leptin receptor, NPY, and AgRP mRNA expression. Am J Physiol Endocrinol Metab 282:E1352–E1359
65. Buettner R, Parhofer KG, Woenckhaus M et al (2006) Defining high-fat-diet ratmodels: metabolic andmolecular effects of different fat types. J Mol Endocrinol 36:485–501
66. Surwit RS, Feinglos MN, Rodin J et al (1995) Differential effects of fat and sucrose on the development of obesity and diabetes in C57BL/6J and A/J mice. Metabolism 44:645–651
67. Chicco A, D'Alessandro ME, Karabatas L et al (2003) Muscle lipid metabolism and insulin secretion are altered in insulin-resistant rats fed a high sucrose diet. J Nutr 133:127–133
68. Banks WA, Coon AB, Robinson SM et al (2004) Triglycerides induce leptinresistance at the blood-brain barrier. Diabetes 53:1253–1260
69. Benoit SC, Kemp CJ, Elias CF et al (2009) Palmitic acid mediates hypothalamic insulin resistance by altering PKC-theta subcellular localization in rodents. J Clin Invest 119:2577–2589
70. Clegg DJ, Gotoh K, Kemp C et al (2011) Consumption of a high-fat diet induces central insulin resistance independent of adiposity. Physiol Behav 103:10–16
71. Hariri N, Thibault L (2010) High-fat diet-induced obesity in animal models. Nutr Res Rev 23:270–299

72. Woods SC, D'Alessio DA, Tso P et al (2004) Consumption of a high-fat diet alters the homeostatic regulation of energy balance. Physiol Behav 83:573–578
73. West DB, Boozer CN, Moody DL et al (1992) Dietary obesity in nine inbred mouse strains. Am J Physiol 262:R1025-R1032
74. Collins S, Martin TL, Surwit RS et al (2004) Genetic vulnerability to diet-induced obesity in the C57BL/6J mouse: physiological and molecular characteristics. Physiol Behav 81:243–248
75. Rossmeisl M, Rim JS, Koza RA et al (2003) Variation in type 2 diabetes-related traits in mouse strains susceptible to diet-induced obesity. Diabetes 52:1958–1966
76. Srinivasan K, Ramarao P (2007) Animal models in type 2 diabetes research: an overview. Indian J Med Res 125:451–472
77. Perez C, Fanizza LJ, Sclafani A (1999) Flavor preferences conditioned by intragastric nutrient infusions in rats fed chow or a cafeteria diet. Appetite 32:155–170
78. Rogers PJ, Blundell JE (1984) Meal patterns and food selection during the development of obesity in rats fed a cafeteria diet. Neurosci Biobehav Rev 8:441–453
79. Rothwell NJ, Stock MJ (1979) Combined effects of cafeteria and tube-feeding on energy balance in the rat. Proc Nutr Soc 38:5A
80. Challis BG, Coll AP, Yeo GS et al (2004) Mice lacking pro-opiomelanocortin are sensitive to high-fat feeding but respond normally to the acute anorectic effects of peptide-YY(3-36). Proc Natl Acad Sci U S A 101:4695–4700
81. Yaswen L, Diehl N, Brennan MB et al (1999) Obesity in the mouse model of proopiomelanocortin deficiency responds to peripheral melanocortin. Nat Med 5:1066–1070
82. O'Rahilly S (2009) Human genetics illuminates the paths to metabolic disease. Nature 462:307–314
83. Huszar D, Lynch CA, Fairchild-Huntress V et al (1997) Targeted disruption of the melanocortin-4 receptor results in obesity in mice. Cell 88:131–141
84. Ste Marie LM, Miura GI, Marsh DJ et al (2000) A metabolic defect promotes obesity in mice lacking melanocortin-4 receptors. Proc Natl Acad Sci U S A 97:12339–12344
85. Mul JD, van Boxtel R, Bergen DJ et al (2012) Melanocortin receptor 4 deficiency affects body weight regulation, grooming behavior, and substrate preference in the rat. Obesity 20:612–621
86. Butler AA, Kesterson RA, Khong K et al (2000) A unique metabolic syndrome causes obesity in the melanocortin-3 receptor-deficient mouse. Endocrinology 141:3518–3521
87. Brobeck JR, Tepperman J, Long CNH (1943) Experimental hypothalamic hyperphagia in the Albino rat. Yale J Biol Med 15:831–853
88. Brooks CM, Lockwood RA, Wiggins ML (1946) A study of the effect of hypothalamic lesions on the eating habits of the albino rat. Am J Physiol 147:735–741
89. Cox JE, Powley TL (1981) Prior vagotomy blocks VMH obesity in pair-fed rats. Am J Physiol 240:E573–E583
90. King BM (1991) Ventromedial hypothalamic obesity: a reexamination of the irritative hypothesis. Neurosci Biobehav Rev 15:341–347
91. King BM (2006) The rise, fall, and resurrection of the ventromedial hypothalamus in the regulation of feeding behavior and body weight. Physiol Behav 87:221–244
92. Penicaud L, Larue-Achagiotis C, Le Magnen J (1983) Endocrine basis for weight gain after fasting or VMH lesion in rats. Am J Physiol 245:E246–E252
93. Dawson R, Wallace DR, Gabriel SM (1989) A pharmacological analysis of food intake regulation in rats treated neonatally with Monosodium L-Glutamate (MSG). Pharmacol Biochem Behav 32:391–398
94. Hirata AE, Andrade IS, Vaskevicius P et al (1997) Monosodium glutamate (MSG)-obese rats develop glucose intolerance and insulin resistance to peripheral glucose uptake. Braz J Med Biol Res 30:671–674
95. Morrison JFB, Shehab S, Sheen R et al (2008) Sensory and autonomic nerve changes in the monosodium glutamatetreated rat: a model of type II diabetes. Exp Physiol 93:213–222

96. Nemeroff CB, Lipton MA, Kizer JS (1978) Models of neuroendocrine regulation: use of monosodium glutamate as an investigational tool. Dev Neurosci 1:102–109
97. Bergen HT, Mizuno TM, Taylor J et al (1998) Hyperphagia and weight gain after gold-thioglucose: relation to hypothalamic neuropeptide Y and proopiomelanocortin. Endocrinology 139:4483–4488
98. Young JK (1992) Hypothalamic lesions increase neuronal immunoreactivity for neuropeptide Y. Brain Res Bull 29:375–380
99. Young JK, McKenzie JC, Brady LS et al (1994) Hypothalamic lesions increase levels of neuropeptide Y mRNA in the arcuate nucleus of mice. Neurosci Lett 165:13–17

Chapter 11
From Homo Obesus to Homo Diabesus: Neuroadipology Insight

George N. Chaldakov, Luigi Aloe, Anton B. Tonchev and Marco Fiore

Introduction

In 2000 Astrup and Finer wrote [1]: "Since type 2 diabetes is obesity dependent, and obesity is the main aetiological cause of type 2 diabetes, we propose the term 'diabesity' should be adopted" (see also [2]). Obesity is most prevalent disease in the world. In 2005, 800 million people were overweight (BMI 25.0–29.9 kg/m^2) and 400 million were obese (BMI over 30 kg/m^2). Although the pathogenesis is not yet completely understood, there is now solid evidence that type 2 (non-insulin dependent) diabetes mellitus is a disease strongly associated with the obese man (*Homo obesus*) [3]. Obesity predisposes to type 2 diabetes and is largely responsible for its current pandemic feature predicting to double the number of diabetic people worldwide within the period of 30 years, from 150 million in 1995 to over 300 million in 2025. Thus *diabesity* has been moving to centre stage being one of the most challenging biomedical and social threats. Accordingly, the term *Homo diabesus* was recently introduced [4].

G. N. Chaldakov (✉) · A. B. Tonchev
Laboratory of Cell Biology, Department of Anatomy and Histology, Medical University, BG-9002 Varna, Bulgaria
e-mail: chaldakov@yahoo.com

L. Aloe · M. Fiore
Institute of Cell Biology and Neurobiology, National Research Council (CNR), Rome, Italy

C. Nóbrega, R. Rodriguez-López (eds.), *Molecular Mechanisms Underpinning the Development of Obesity*, DOI 10.1007/978-3-319-12766-8_11

Adipobiology: The Renaissance of a Tissue Marked by Three Paradigm Shifts

One of biggest recent achievements in studying cardiovascular (atherosclerosis and hypertension) and metabolic (obesity, type 2 diabetes, metabolic syndrome, and Alzheimer's disease which is recently viewed as type 3 diabetes, see below) diseases is associated with the "rediscovery" of a neglected tissue, the adipose tissue.

In 1962 Thomas S. Kuhn published his book *The Structure of Scientific Revolutions* (1st edition, University of Chicago Press, Chicago, USA). Its publication was a landmark event in history and philosophy of scientific knowledge (epistemology). Kuhn challenged the then prevailing view of "normal science" which was viewed as "development-by-accumulation" of accepted facts and concepts leading—most oftenly—to *epistemological paralysis*, we dubbed it neophobia. Kuhn argued for a model in which a period of such conceptual continuity in normal science were interrupted by a period of revolutionary science leading to a new paradigm, an event he designated *paradigm shift*.

At epistemological level, the adipose tissue has undergone three major paradigm shifts in last 20 years. This rise it above the horizon to take center stage in so many syndromes and diseases that it leaves most scientists and medical doctors astonished.

The first paradigm shift was: while considered as passive storage-release of lipids by most cell biologists and pathologists for a long period of time, adipose tissue is now considered the biggest endocrine and paracrine organ of the human body (Table 11.1). The discovery of leptin, an adipose-secreted hormone/cytokine, by Jeffrey Friedman and colleagues in 1994 marked this revolutionary event (Table 11.2). Here the pioneering contribution of Douglas Coleman (1931–2014) has to be acknowledged. His work established the first clues to a genetic component in obesity. In the 1970s, Coleman conducted a series of experiments that led him to propose the existence of a *satiety factor* that would account for obesity and type 2 diabetes among certain laboratory mice.

Table 11.1 A paradigm shift: never before has adipose tissue been so active

From
Adipose tissue is a lipid/energy storage involved in obesity
To
Adipose tissue is an endocrine and paracrine organ
Adipose tissue is a neuroedocrine organ
Adipose tissue is a steroidogenic organ
Adipose tissue is an immune organ
Adipose tissue is a source of and target for inflammatory mediators
Adipose tissue produces all components of rennin-angiotensin system
Adipose tissue is thus involved in numerous diseases beyond obesity

Table 11.2 Adipose-brain talk: examples of mediators in leptin signaling

Anorexigenic pathway ↑	Orexigenic pathway ↓
Proopiomelanocortins	Neuropeptide tyrosine (NPY)
Melanocortin 4	*Agouti*-related protein
α-melanocyte stimulating hormone	Endocannabinoids
Brain-derived neurotrophic factor	

The second paradigm shift derived from the study of Jeffrey Bell and colleagues [5] who have scanned nearly 800 people with magnetic resonance imaging (MRI) technique, aimed at obtaining map of WAT. The authors demonstrated that as many as 45 % of women and nearly 60 % of men scanned have normal scores of the body mass index (BMI, 20–25 kg/m^2). These people are thin outside (TO), while actually have excessive levels of internal adipose tissue—they are fat inside (FI), hence TOFI phenotype of body fatness. Noteworthy, TOFI phenotype was also found among people who are professional models. TOFI may thus be considered a specific, "invisible" expression of both *Homo obesus* [3] and *Homo diabesus* [4].

The third paradigm shift features the increasing significance of brown adipose tissue in health and disease (see below).

Accumulation of adipose tissue in the visceral and subcutaneous abdominal tissue, also around almost all internal organs, is a major risk factor for the development of numerous disorders including diabesity and related cardiometabolic diseases. Of note, *metaflammation* (metabolically-induced inflammation) has emerged as a pivotal process involved in the clustering of those disorders [6].

Adipose tissue is very plastic tissue, being constantly remodeled along with weight gain and weight loss. It is a dynamic cellular and extracellular matrix assembly composed of adipocytes, fibroblasts, immune cells and matrix components, also rich in sympathetic nerve fibers, blood vessels, and stem cells. There are two major subtypes of adipose tissue, white adipose tissue (WAT) and brown adipose tissue (BAT).

By sending and receiving different types of protein and non-protein signals, adipose tissue communicates with many organs in the body (Fig. 11.1), thus contributing to the control of energy, lipid and glucose homeostasis as well as inflammation, immunity, learning and memory among many other biological functions.

In human body, while WAT stores energy, BAT has the ability to dissipate energy by producing heat. BAT-mediated increase in energy expenditure is realized by uncoupling respiration from ATP synthesis, via uncoupling protein 1 (UCP1), which is expressed in brown adipocytes, thus generating heat, a process known as adaptive thermogenesis. Animal studies have shown that activation of BAT counteracts diet-induced weight gain and related disorders such as type 2 diabetes and metabolic syndrome; it may also be the case for humans [7]. Recently, the knowledge about WAT and BAT were enriched with their relatives, namely *brite* (brown in white) and *bruscle* (brown in skeletal muscle) adipocytes [8]. Hence, brown adipobiology is emerging as a new challenge in biomedicine.

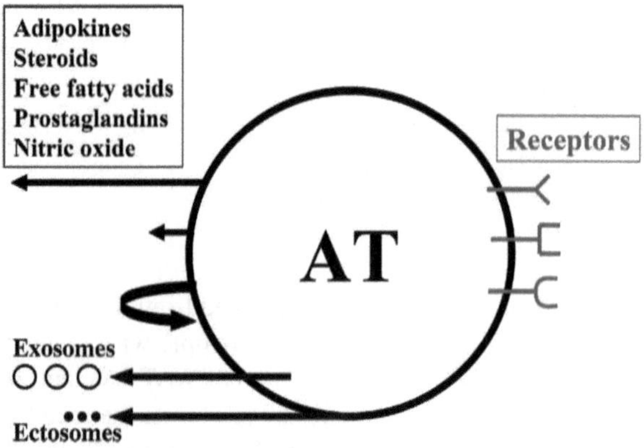

Fig. 11.1 A drawing illustrating both secretory and receptor nature of adipose tissue (*AT*) cells. At the secretory level, AT-derived signaling molecules communicate via multiple pathways, such as endocrine (*arrows 1, 4 and 5, from top to bottom*), paracrine (*arrow 2*) and autocrine (*arrow 3, curved*). Also depicted is that AT cells express receptors for various ligands

In effect, such an adipocentric approach has revealed that while BAT is major thermogenic organ, WAT is the body's largest endocrine and paracrine organ producing multiple signaling proteins collectively termed adipokines [9–12], with nerve growth factor (NGF) and brain-derived neurotrophic factor (*BDNF*) being also produced from adipose tissue [13].

Multiple Life of Neurotrophins and Adipokines

At the end of the nineteenth century it was envisaged by Santiago Ramon y Cajal but has not been proved that the nerves require trophic support. By a rare combination of scientific reasoning and intuition, Rita Levi-Montalcini (1909–2012) obtained the proof in the early 1950s in Saint Louis, MO, USA, where the first cell growth factor, namely NGF, was discovered, and 35 years later awarded Nobel Prize 1986 [14]. Data of NGF have been embodied in a conceptual framework well known now as neurotrophic (nerve-effector interaction) theory. It reveals a pivotal role of effector (target) cells in the control of neuronal differentiation, survival and function via production of NGF and other neurotrophic factors.

The neurotrophin family of proteins consisted of NGF, pro-NGF, brain-derived neurotrophic factor (BDNF), pro-BDNF, neurotrophin-3 (NT-3), NT-4/5, NT-6, and NT-7. Neurotrophins mediate their effects via ligation of (i) panneurotrophin receptor, $p75^{NTR}$, and (ii) receptor tyrosine kinase (tropomyosin-related kinase) (Trk), namely, TrkA (for NGF), Trk B (for BDNF and NT-4/5), and TrkC (for NT-3) [14, 15].

The past three decades has witnessed a number of breakthroughs in the study on Rita Levi-Montalcini's NGF. Studies have revealed that the neurotrophins NGF and BDNF are not only stimulating for nerve growth and survival, but also exert trophic effects over (i) immune cells, acting as immunotrophins, (ii) keratinocytes, enterocytes, prostate and breast epithelial cells, acting as epitheliotrophins, and (iii) endothelial cells, acting as angiogenic factors (reviewed in [15]).

Metabotrophic Factors and Cardiometabolic Disease

In 2003 the *NGF-ome* was enriched with one more expression, that is, metabotrophic action on glucose, lipid, energy, pancreatic beta cell and cardiovascular homeostasis, and thus, together with BDNF, designated metabotrophic factors (MTF) or metabotrophins (from Greek *metabole*, and *trophe*, nutrition, means "nutritious for metabolism") (reviewed in [15, 16]; see also [17]). Onward, the proof-of-hypothesis was based on results demonstrating that the circulating and/or tissue levels of NGF and BDNF are (commonly) decreased in atherosclerosis and metabolic syndrome [18], type 2 diabetes [19, 20], depression and other psychiatric diseases [15], and in Alzheimer's disease which nowadays is considered as type 3 diabetes [21, 22].

Neuroadipology

Life of multicellular organisms requires the interaction between cells of nervous and other systems. One of the biggest recent achievements of neurobiology and adipobiology is the study on neurotrophic factors (e.g. NGF, BDNF) and adipokines (e.g. leptin, adiponectin) respectively.

As often occurs, the framework of an initial concept of the physiological role of a newly discovered molecules extends in the light of emerging findings. This was also the case with neurotrophic factors and adipokines. For instance, during some 30 years after NGF discovery, there have been few reasons given to indicate that it acts on non-neuronal cells. Thus, it was remarkable when Aloe and Levi-Montalcini have discovered in 1977 that the treatment of newborn rats with NGF caused a systemic increase in the number of mast cells. This seminal finding paved the road of a novel research field, neuroimmunology ([23] and references therein).

As indicated above [9–12] the adipose tissue is a dynamic endocrine and paracrine organ producing a large number of adipokines. Some of them mediate the cross-talk between adipose tissue and hypothalamus in regulating food intake and energy homoeostasis (Table 11.2 for leptin). However, the hypothalamus is not the only brain target for leptin, and food intake is not the only biological effect of this adipokine. Rather, some adipokines support various cognitive functions and exert neurotrophic activity. Current data of adipose-derived neuroendocrine and neurotrophic factors are summarized in Tables 11.3 and 11.4. This raises an intriguing question as to whether adipose tissue might be a peripheral counterpart of hypothalamus-hypophysis (see also [41]). Cumulatively, linking neurobiology and adipobiology resulted in neuroadipology, a novel component of neuroendocrinology [24].

In an attempt to "close" the metabotrophic "loop" in cardiometabolic disease, we have measured circulating levels of NGF and BDNF in patients with acute coronary syndromes, and found they are significantly reduced [25, 26]. Another study revealed altered levels of NGF in pancreas and brain in streptozotocin-induced

Table 11.3 Selected list of adipose-derived neuroendocrine factors

Neuropeptides	Hypothalamic factors
Neuropeptide tyrosine (NPY)	Mineralocorticoid-releasing factors
Substance P	Corticotropin-releasing hormone (CRH)
Calcitonin gene-related peptide	Stresscopin, urocortin (CRH-like peptides)
Agouti protein	
Adrenomedullin	
Somatostatin	
Kisspeptin	
Neuromedin B	
Neurotensin	
Apelin	
Nesfatin-1	

Table 11.4 Selected list of adipose-derived neurotrophic factors

Leptin
Nerve growth factor
Brain-derived neurotrophic factor
Angiopoietin-1
Vascular endothelial growth factor
Ciliary neurotrophic factor
Glial cell line-derived neurotrophic factor
Steroids
Metallothioneins

diabetes [27]. Recently, it was demonstrated that in response to experimental stress or diabetes, the amount of NGF and BDNF was altered both in WAT and BAT (Figs. 11.2 and 11.3; [13]).

Conclusion and Perspectives

Examples for proof-of-metabotrophic concept derived from other laboratories include: (i) pancreatic beta cells secrete NGF and express its TrkA receptor, findings being implicated in the pathogenesis of diabetes mellitus [28], and (ii) mutations affecting *Bdnf* gene (encoding BDNF) in mice or *Ntr2k2* gene (encoding the high-affinity BDNF receptor TrkB) in patients are associated with hyperphagia and severe obesity (15 and references therein). Selective list of metabotrophins (Table 11.5) and metabotrophic effects of NGF and BDNF (Table 11.6) are shown.

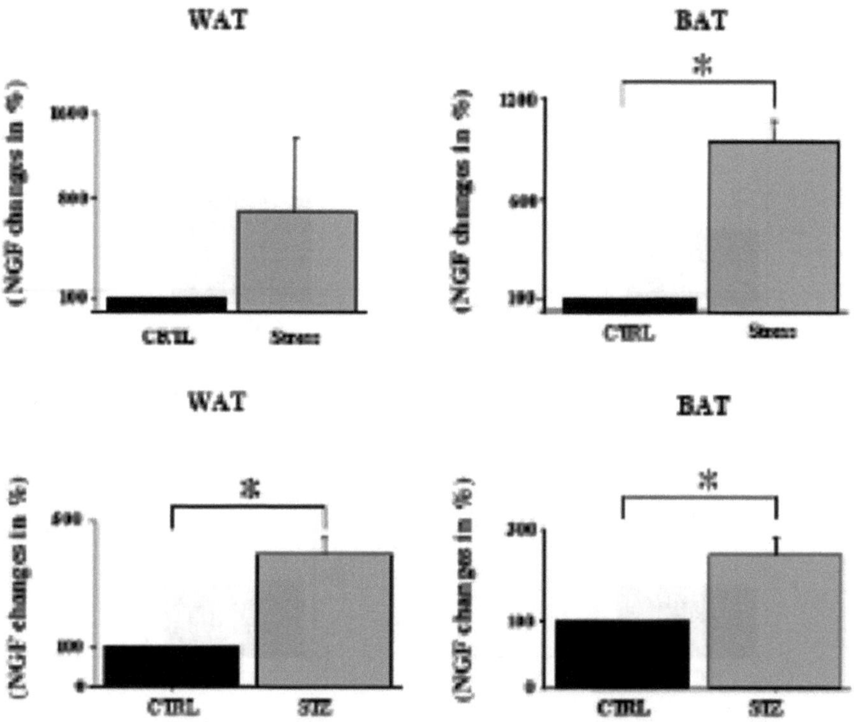

Fig. 11.2 Changes in the amount of nerve growth factor (*NGF*) in white adipose tissue (*WAT*) and brown adipose tissue (*BAT*) of controls compared to the concentration of NGF in stressed mice (*Stress*) and streptozotocin-induced diabetic rats (*STZ*), expressed as percentage of controls. Note the enhanced presence of NGF in WAT and BAT in stressed mice as well as diabetic rats. The vertical lines in the figure indicate pooled S.E.M. derived from appropriate error mean square in the ANOVA. * Significant differences between groups ($p < 0.05$)

In this context, the recent discovery of (i) humanin, a mitochondria-derived peptide expressing neuro-metabotrophic effect [29, 30], and (ii) irisin, a myokine/adipokine involved in the browning of WAT [31, 32], may lead to a novel approach in the therapy of *Homo diabesus*. It may open new windows for the search of *exogenous* MTF, such as (i) small molecules boosting secretory and/or signaling pathways of MTF [15], and (ii) incretin mimetics and receptor agonists, because the insulinotropic hormone glucagon-like peptide-1 (GLP-1) and exendin-4, a GLP-1 receptor agonist, exert neuro-metabotrophic effect [33, 34].

Further, the present description suggests that understanding the precise role of MTF in the origin of *Homo diabesus* may lead to potential new therapies for diabesity and related diseases including Alzheimer's disease (AD). Noteworthy, the use of transcript clustering to identify molecular mechanisms contributing to early stages of AD in mice identified changes to the insulin signaling pathway including the down-regulation of insulin receptor substrate 4 (Irs4) as an early event in AD [46]. Insulin signaling is strongly associated with diabesity, which have recently been identified as possible risk factors for AD [30, 34, 47–52, also see 53, 54].

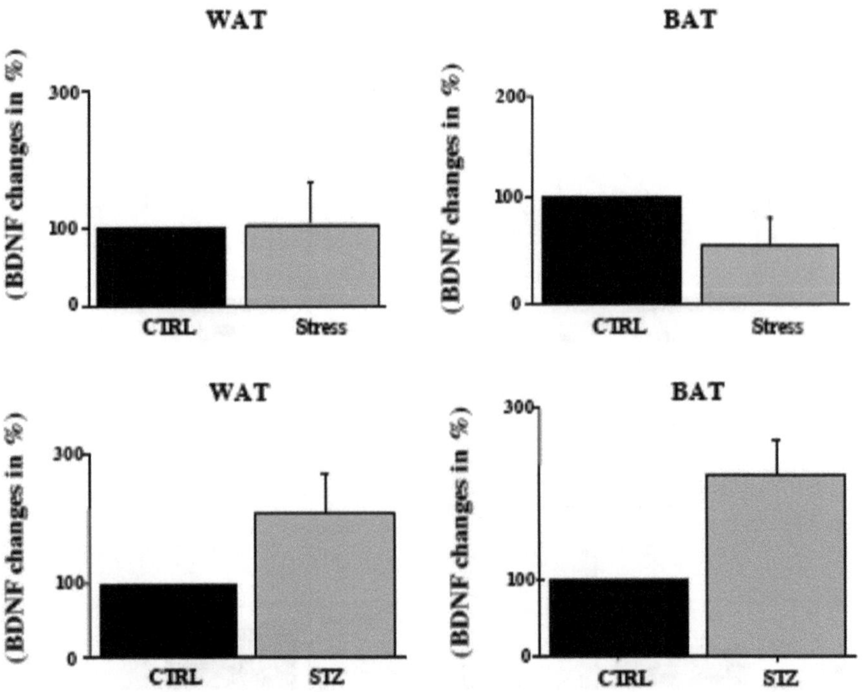

Fig. 11.3 Changes in the amount of brain-derived neurotrophic factor (*BDNF*) in epicardial white adipose tissue (*WAT*) and brown adipose tissue (*BAT*) of controls compared to the concentration of BDNF in stressed mice (*Stress*) and streptozotocin-induced diabetic rats (*STZ*), expressed as percentage of controls. The vertical lines in the figure indicate pooled S.E.M. derived from appropriate error mean square in the ANOVA

Table 11.5 Selected list of endogenous metabotrophic factors. (Modified from [4]. For references see the text, also [35–45])

Secretory proteins	Intracellular proteins
Nerve growth factor, Brain-derived neurotrophic factor	Sitruins, PPAR-gamma, Uncoupling protein-1 (UCP-1)
Ciliary neurtrophic factor, Neuron-derived neurotrophic factor	
Adiponectin, Irisin, Humanin, Omentin, Chemerin, Apelin, Otopetrin 1	
Interleukin-10, Metalothionein-I,-II, Glucagon-like peptide-1	

Table 11.6 Metabotrophic effects of NGF and BDNF. (Modified from [15]. For references see the text, also [39–40])

NGF shares homology with proinsulin
NGF/BDNF are produced by pancreatic beta cells and exert insulinotropic effect
NGF/BDNF are trophic factors for pancreatic beta cells, also improve beta cell transplantation
NGF up-regulates expression of LDL receptor-related protein
NGF up-regulates expression of PPARgamma
NGF inhibits glucose-induced down-regulation of caveolin-1
NGF improves skin and corneal wound healing
NGF may improve vascular (atheroma) wound healing
NGF rescues silent myocardial ischemia in diabetes mellitus
NGF improves diabetic erectile dysfunction
NGF and BDNF suppress food intake
Healthy lifestyle increases brain and/or circulating levels of NGF/BDNF
Atherogenic diet decreases brain BDNF levels
BDNF-deficient mice develop abnormalities similar to the metabolic syndrome
BDNF improves cognitive processes

Coda

In 1999 Albee Messing published in *Hepatology* (29: 602–603) an editorial entitled "Nestin in the liver—lessons from the brain." He wrote: "Most neuroscientists manage to get through each day without thinking of the liver even once … but I think that is about to change." This may also be the case for adipose tissue. Future *dance round in* neuroadipology of diabesity might therefore cultivate a neuro-metabotrophic thinking about how we can make MTF secretion and signaling work for the improvement of physical and mental quality of life of *Homo diabesus*.

References

1. Astrup A, Finer N (2000) Redefining type 2 diabetes: 'diabesity' or 'obesity dependent diabetes mellitus'? Obes Rev 1:57–59
2. Kurland IJ, Accili D, Burant C, Fischer SM, Kahn BB, Newgard CB et al (2013) Application of combined omics platforms to accelerate biomedical discovery in diabesity. Ann N Y Acad Sci 1287:1–16
3. Chaldakov GN, Fiore M, Tonchev AB, Dimitrov D, Pancheva R, Rančič G, Aloe L (2007) *Homo obesus*: a metabotrophin-deficient species. Pharmacology and nutrition insight. Curr Pharm Des 13:2176–2179
4. Aloe L, Tonchev AB, Fiore M, Chaldakov GN (2013) *Homo diabesus*: involvement of metabotrophic factors. Adipobiology 5:45–49

5. Louise TE, Saeed N, Hajnal JV, Brynes A, Goldstone AP, Frost G et al (1998) Magnetic resonance imaging of total body fat. J Appl Physiol 85:1778–1785
6. Jin C, Flavell RA (2013) Innate sensors of pathogen and stress: linking inflammation to obesity. J Allergy Clin Immunol 132:287–294
7. Sacks H, Symonds ME (2013) Anatomical locations of human brown adipose tissue functional relevance and implications in obesity and type 2 diabetes. Diabetes 62:1783–1790
8. Giralt M, Villarrova F (2013) White, brown, beige/brite: different adipose cells for different functions? Endocrinology 154:2992–3000
9. Chaldakov GN, Stankulov IS, Hristova M, Ghenev PI (2003) Adipobiology of disease: adipokines and adipokine-targeted pharmacology. Curr Pharm Des 9:1023–1031
10. Chaldakov GN, Tunçel N, Beltowski J, Fiore M, Rancić G, Tonchev A et al (2012) Adipoparacrinology: an emerging field in biomedical research. Balkan Med J 29:2–9
11. Renes J, Mariman E (2013) Application of proteomics technology in adipocyte biology. Mol Biosyst 9:1076–1091
12. Bouckenooghe T, Sisino G, Aurientis S, Chinetti-Gbaguidi G, Kerr-Conte J, Staels B et al (2014) Adipose tissue macrophages (ATM) of obese patients are releasing increased levels of prolactin during an inflammatory challenge: a role for prolactin in diabesity? Biochim Biophys Acta 1842:584–593
13. Sornelli F, Fiore M, Chaldakov GN, Aloe L (2009) Adipose tissue-derived nerve growth factor and brain-derived neurotrophic factor: results from experimental stress and diabetes. Gen Physiol Biophys 28:179–183
14. Levi-Montalcini R (1987) The nerve growth factor 35 years later. Science 237:1154–1162
15. Yanev S, Aloe L, Fiore F, Chaldakov GN (2013) Neurotrophic and metabotrophic potential of nerve growth factor and brain-derived neurotrophic factor: linking cardiometabolic and neuropsychiatric diseases. World J Pharmacol 2:92–99
16. Chaldakov GN (2011) The metabotrophic NGF and BDNF: an emerging concept. Arch Ital Biol 149:257–263
17. Gomez-Pinilla F, Vaynman S, Ying Z (2008) Brain-derived neurotrophic factor functions as a metabotrophin to mediate the effects of exercise on cognition. Eur J Neurosci 28:2278–2287
18. Chaldakov GN, Fiore M, Stankulov IS, Manni L, Hristova MG, Antonelli A et al (2004) Neurotrophin presence in human coronary atherosclerosis and metabolic syndrome: a role for NGF and BDNF in cardiovascular disease? Prog Brain Res 146:279–289
19. Krabbe KS, Nielsen AR, Krogh-Madsen R, Plomgaard P, Rasmussen P, Erikstrup C et al (2007) Brain-derived neurotrophic factor (BDNF) and type 2 diabetes. Diabetologia 50:431–438
20. Yamanaka M, Itakura Y, Ono-Kishino M, Tsuchida A, Nakagawa T, Taiji M (2008) Intermittent administration of brain-derived neurotrophic factor (BDNF) ameliorates glucose metabolism and prevents pancreatic exhaustion in diabetic mice. J Biosci Bioeng 105:395–402
21. de la Monte SM, Wands JR (2008) Alzheimer's disease is type 3 diabetes-evidence reviewed. J Diabetes Sci Technol 2:1101–1113
22. Dar TA, Sheikh IA, Ganie SA, Ali R, Singh LR, Gan SH et al (2014) Molecular linkages between diabetes and Alzheimer's disease: current scenario and future prospects. CNS Neurol Disord Drug Targets 13(2):290–298
23. Chaldakov GN, Fiore M, Ghenev PI, Beltowski J, Rancic G, Tunçel N, Aloe L (2014) Triactome: neuro-immune-adipose interactions. Implication in vascular biology. Front Immunol 5:130
24. Chaldakov GN, Fiore M, Tonchev AB, Aloe L (2010) Neuroadipology: a novel component of neuroendocrinology. Cell Biol Int 34:1051–1053
25. Manni L, Nikolova V, Vyagova D, Chaldakov GN, Aloe L (2005) Reduced plasma levels of NGF and BDNF in patients with acute coronary syndromes. Int J Cardiol 102:169–171
26. Ejiri J, Inoue N, Kobayashi S, Shiraki R, Otsui K, Honjo T et al (2005) Possible role of brain-derived neurotrophic factor in the pathogenesis of coronary artery disease. Circulation 112:2114–2120

27. Sposato V, Manni L, Chaldakov GN, Aloe L (2007) Streptozotocin-induced diabetes is associated with changes in NGF levels in pancreas and brain. Arch Ital Biol 145:87–97
28. Larrieta ME, Vital P, Mendoza-Rodriguez A, Cerbón M, Hiriart M (2006) Nerve growth factor increases in pancreatic beta cells after streptozotocin-induced damage in rats. Exp Biol Med (Maywood) 231:396–402
29. Hoang PT, Park P, Cobb LJ, Paharkova-Vatchkova V, Hakimi M, Cohen P et al (2010) The neurosurvival factor Humanin inhibits beta-cell apoptosis via signal transducer and activator of transcription 3 activation and delays and ameliorates diabetes in nonobese diabetic mice. Metabolism 59:343–349
30. Mahboobi H, Golmirzaei J, Gan SH, Jalalian M, Jalalian M (2014) Humanin: a possible linkage between Alzheimer's disease and type 2 diabetes. CNS Neurol Disord Drug Targets 13(3):543–552 (2013 Dec 22. [Epub ahead of print])
31. Novelle MG, Contreras C, Romero-Picó A, López M, Diéguez C (2013) Irisin, two years later. Int J Endocrinol 2013:746281
32. Spiegelman BM (2013) Banting lecture 2012: regulation of adipogenesis: toward new therapeutics for metabolic disease. Diabetes 62:1774–1782
33. Perry T, Lahiri DK, Chen D, Zhou J, Shaw KT, Egan JM et al (2002) A novel neurotrophic property of glucagon-like peptide 1: a promoter of nerve growth factor-mediated differentiation in PC12 cells. J Pharmacol Exp Ther 300:958–966
34. Li L (2007) Is glucagon-like peptide-1, an agent treating diabetes, a new hope for Alzheimer's disease? Neurosci Bull 23:58–65
35. Iacobellis G, Di Gioia C, Petramala L, Chiappetta C, Serra V, Zinnamosca L et al (2013) Brown fat expresses adiponectin in humans. Int J Endocrinol 2013:126751
36. Tan BK, Adya R, Randeva HS (2010) Omentin: a novel link between inflammation, diabesity, and cardiovascular disease. Trends Cardiovasc Med 20:143–148
37. Castan-Laurell I, Dray C, Attané C, Duparc T, Knauf C, Valet P (2011) Apelin, diabetes, and obesity. Endocr 40:1–9
38. Wang GX, Cho KW, Uhm M, Hu CR, Li S, Cozacov Z et al (2014) Otopetrin 1 protects mice from obesity-associated metabolic dysfunction through attenuating adipose tissue inflammation. Diabetes 63(4):1340–1352
39. Karatzas A, Katsanos K, Lilis I, Papadaki H, Kitrou P, Lecht S et al (2013) NGF promotes hemodynamic recovery in a rabbit hindlimb ischemic model through trkA- and VEGFR2-dependent pathways. J Cardiovasc Pharmacol 62:270–277
40. Aloe L, Tirassa P, Lambiase A (2008) The topical application of nerve growth factor as a pharmacological tool for human corneal and skin ulcers. Pharmacol Res 57:253–258
41. Schäffler A, Schölmerich J, Buechler C (2006) The role of "adipotrophins" and the clinical importance of a potential hypothalamic-pituitary-adipose axis. Nat Clin Pract Endocrinol Metab 2:374–383
42. Hausman GJ, Barb CR, Dean RG (2008) Patterns of gene expression in pig adipose tissue: insulin-like growth factor system proteins, neuropeptide Y (NPY), NPY receptors, neurotrophic factors and other secreted factors. Domest Anim Endocrinol 35:24–34
43. Rao AA (2013) Views and opinion on BDNF as a target for diabetic cognitive dysfunction. Bioinformation 9:551–554
44. Meek TH, Wisse BE, Thaler JP, Guyenet SJ, Matsen ME, Fischer JD et al (2013) BDNF action in the brain attenuates diabetic hyperglycemia via insulin-independent inhibition of hepatic glucose production. Diabetes 62:1512–1518
45. Byerly MS, Swanson RD, Semsarzadeh NN, McCulloh PS, Kwon K, Aja S et al (2013) Identification of hypothalamic neuron-derived neurotrophic factor as a novel factor modulating appetite. Am J Physiol Regul Integr Comp Physiol 304:R1085–R1095
46. Jackson HM, Soto I, Graham LC, Carter GW, Howell GR (2013) Clustering of transcriptional profiles identifies changes to insulin signaling as an early event in a mouse model of Alzheimer's disease. BMC Genomics 14:831
47. Luchsinger JA, Mayeux R (2007) Adiposity and Alzheimer's disease. Curr Alzheimer Res 4:127–134

48. Naderali EK, Ratcliffe SH, Dale MC (2009) Review: obesity and Alzheimer's disease: a link between body weight and cognitive function in old age. Am J Alzheimers Dis Other Demen 24:445–449
49. Frisardi V, Solfrizzi V, Seripa D, Capurso C, Santamato A, Sancarlo D et al (2010) Metabolic-cognitive syndrome: a cross-talk between metabolic syndrome and Alzheimer's disease. Ageing Res Rev 9:399–417
50. O'Neill C, Kiely AP, Coakley MF, Manning S, Long-Smith CM (2012) Insulin and IGF-1 signalling: longevity, protein homoeostasis and Alzheimer's disease. Biochem Soc Trans 40:721–727
51. de la Monte SM (2012) Brain insulin resistance and deficiency as therapeutic targets in Alzheimer's disease. Curr Alzheimer Res 9:35–66
52. Hildreth KL, Van Pelt RE, Schwartz RS (2012) Obesity, insulin resistance, and Alzheimer's disease. Obesity 20:1549–1557
53. Kimmel J, Sugaya K (2014) Nanog overexpression increases neural marker expression in adipose derived stem cells. Recent Patents Regener Med 4:69–74
54. Revollo JR, Körner A, Mills KF, Satoh A, Wang T, Garten A et al (2007) Nampt/PBEF/Visfatin regulates insulin secretion in beta cells as a systemic NAD biosynthetic enzyme. Cell Metab 6:363–375

Chapter 12
Obesity and Type 2 Diabetes

Charbel Abi Khalil and Shahrad Taheri

Introduction

The impact of excess adiposity (overweight and obesity) in health and wellbeing has been appreciated throughout history. From an evolutionary standpoint, adiposity could provide a survival advantage in an environment where resources are scarce. In fact, there is a close relationship between adipose tissue stores and reproductive function, which is essential for species survival. Whilst maintaining adipose tissue stores has served survival in previous environments, recent environments that promote excess calorie intake and reduced physical activity have resulted in the current obesity pandemic, which has become one of the greatest challenges to health. The pervasive impact of obesity on health has now been recognised by some organisations designating obesity as a disease condition.

While accurate data is not universally available, the incidence and prevalence of obesity is increasing worldwide [1, 2]. According to the World Health Organization (WHO), more than 1 in 10 adults are obese worldwide [2]. The WHO estimated in 2008 that over 1.4 billion adults aged 20 years and above were overweight, and 300 million men and 200 million women were obese [2]. Estimates in 2012, put 40 million children under the age of 5 years as overweight or obese. Amongst high-income countries, the United States of America (USA) has the highest prevalence of overweight and obesity (30% of the population) with a projection for this to rise to 50% of the population by 2030. In USA, obesity accounts for over 20% of health-care expenditure. With increasing levels of overweight and obesity, the prevalence

S. Taheri (✉)
Department of Medicine, Weill Cornell Medical College—Qatar, Qatar Foundation—Education City, PO Box 24144, Doha, Qatar
e-mail: szt2004@qatar-med.cornell.edu

C. Abi Khalil
Department of Genetic Medicine and Department of Medicine, Weill Cornell Medical College—Qatar, Qatar Foundation—Education City, PO Box 24144, Doha, Qatar

C. Nóbrega, R. Rodriguez-López (eds.), *Molecular Mechanisms Underpinning the Development of Obesity*, DOI 10.1007/978-3-319-12766-8_12

of chronic conditions such as type 2 diabetes mellitus is also increasing. The WHO reports that approximately 347 million people have diabetes worldwide, and that by 2030, diabetes will be the 7th leading cause of death [3]. In USA, obesity is associated with low socioeconomic status and poverty and certain ethnic groups are vulnerable to obesity (e.g. Hispanic and African Americans; [4]. In Middle Eastern countries, particularly those with high gross domestic product, paradoxically, obesity is associated with greater affluence and westernization [1].

Of major concern are the increasing levels of obesity in children and adolescents. The onset of obesity at a young age can result in significant morbidity and mortality. In a cohort of young men ($n=6502$, age 22 years) followed up in Denmark [5], obesity was associated with serious adverse outcomes: 48% of those who were obese (BMI $\geq$30 kg/m^2) had diabetes, cardiovascular disease (CVD), or venous thromboembolism, or had died before the age of 55 years. Compared to normal weight participants, the absolute risk difference was 28% (95% CI 19–38%) and the hazard ratio was 3.0 (95% CI 2.3–4.0). In a cohort study of healthy young men ($n=37.674$) [6], there was an independent association between elevated body mass index (BMI) at age 17 years and angiography-proven coronary heart disease ($\beta=1.355$, $P=0.004$).

Type 2 diabetes results in serious micro-and macro-vascular complications. Microvascular complications include nephropathy, retinopathy, and neuropathy. Diabetes is the leading cause of blindness, end-stage renal disease, and non-traumatic amputations. Macrovascular complications include ischaemic heart disease, stroke, and peripheral arterial disease. Type 2 diabetes mellitus associated with obesity at a young age can also be devastating, being associated with a greater burden of micro and macrovascular complications. A Swedish population-based study reported that 39% of young adults with diabetes developed retinopathy at 10 years [7]. Furthermore those with type 2 diabetes had 3 times greater risk for severe retinopathy compared to those with type 1 diabetes. For cardiovascular disease, a study of UK primary care data [8], reported that the adverse cardiovascular risk profiles of the younger aged with diabetes (mean age 33.8 years) group were similar to the older group (mean age 66.9 years). Regarding mortality, in an Australian study [9], the outcomes for 354 type 2 diabetes patients with early onset diabetes (aged 15–30 years) were compared to type 1 diabetes patients. Median follow up was greater than 20 years. Those with type 2 diabetes had almost double the mortality compared to the type 1 diabetes group (11% *vs.* 6%). Furthermore, death in the type 2 diabetes group occurred after about 10 years shorter disease duration than the type 1 diabetes group. There were more cardiovascular deaths (50% *vs.* 30%), and also increased neuropathy, in the type 2 diabetes group.

Obesity has many complications (Fig. 12.1), but dysglycemic disorders are most closely linked to it. It is commonly agreed that excess adiposity (in particular excess visceral adiposity; [10]) results in insulin resistance, which in turn is associated with high insulin secretion. A number of factors promote insulin resistance including pathways influenced by adipocytokines [11–13]. Eventually, high insulin secretion exhausts pancreatic islet beta cell function resulting in insulin dependence. A number of factors will determine the development and progression of type 2 diabetes (Fig. 12.2) with obesity including genetic and epigenetic factors [14, 15] that influence

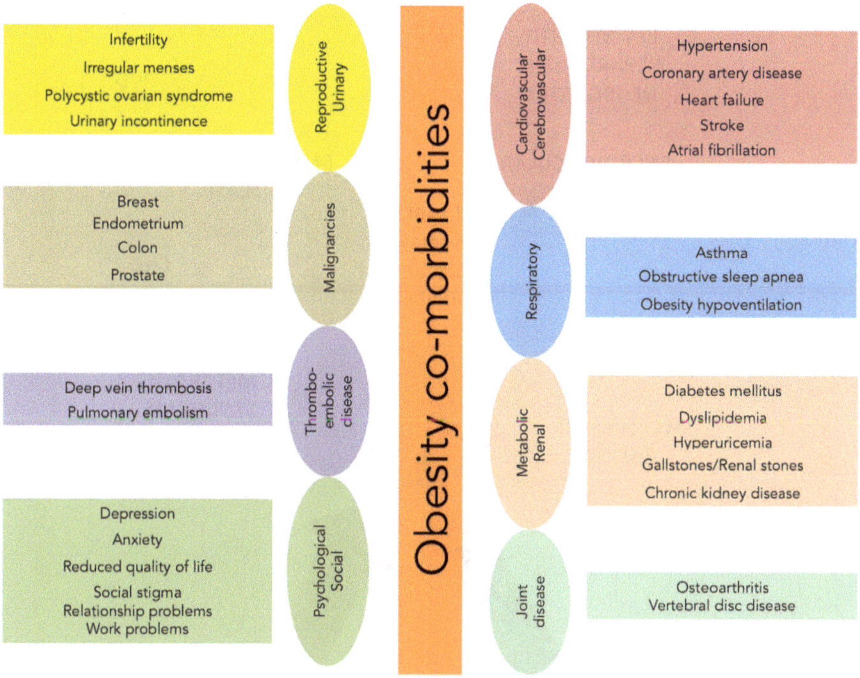

Fig. 12.1 The multiple complications of obesity. Obesity affects multiple organs and systems. The complications of obesity significantly reduce quality of life and increase mortality

Fig. 12.2 Factors associated with diabetes development and progression

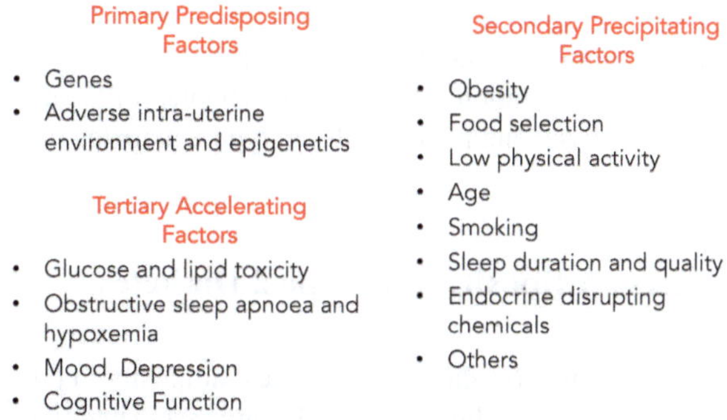

pancreatic beta cell number and function. Abnormalities in the neuro-hormonal axes are also important in the relationship between obesity and type 2 diabetes. Other factors that have recently been recognised to potentially mediate the relationship between obesity and type 2 diabetes [16] including endocrine disrupting chemicals [17] and the gastrointestinal microbiome [18].

This chapter will discuss the relationship between obesity and diabetes and the potential mechanisms involved. The major cause of death in diabetes is cardiovascular disease and the most devastating and financially costly diabetes microvascular

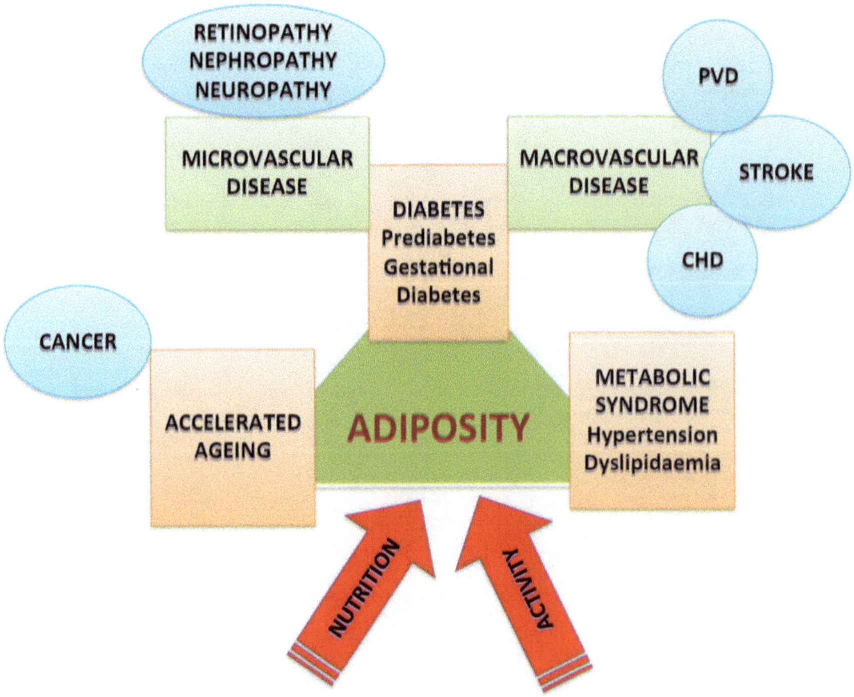

Fig. 12.3 The central role of adiposity in the metabolic syndrome, dysglycaemia, diabetes, and accelerated ageing. *PVD* peripheral vascular disease, *CHD* coronary heart disease. All factors are interconnected and synergistically influence cardiometabolic and other health outcomes

complication is end-stage renal disease requiring renal replacement therapy and renal transplantation. Therefore, the chapter will concentrate on mortality, coronary artery disease, and nephropathy in type 2 diabetes in the context of obesity.

Excess Adiposity and Type 2 Diabetes

There is extensive data linking excess adiposity to type 2 diabetes (Fig. 12.3). Even small increases in body mass index are associated with a significant increase in risk of developing dysglycemia and type 2 diabetes [19]. Several diabetes prevention studies have examined the impact of lifestyle change and weight loss on progression into type 2 diabetes [20, 21]. These have shown that weight loss of 5–10 % and lifestyle change (dietary change and increased physical activity) are associated with significant reductions in the risk of progression into type 2 diabetes [20]. The recent Look AHEAD (Action for Health in Diabetes) study in the USA examined the impact of intensive lifestyle intervention on type 2 diabetes itself with the primary outcome of cardiovascular mortality [22]. While, the study did not find any reduction in cardiovascular mortality with intensive lifestyle intervention and weight loss, there were significant improvements in glycated hemoglobin (HbA1c) and medication use. The Early ACTID study (Early ACTivity In Diabetes) in southwest England examined the impact of dietary and physical activity in early diabetes and observed reductions in HbA1c with reduction in body weight and waist circumference [23].

Recently, there has been great interest in the impact of bariatric surgery on type 2 diabetes [24, 25]. Bariatric surgery results in the most significant weight loss that is more likely to be best sustained [26]. The procedures employed result in varying degrees of weight loss, but those that result in the greatest weight loss are associated with the greatest percentage of patients achieving improvement and remission of diabetes [27]. The STAMPEDE trial [28] reported 3-year outcomes of intensive medical management versus intensive medical management plus bariatric surgery (sleeve gastrectomy and Roux-en-Y gastric bypass) for patients with uncontrolled type 2 diabetes. Weight loss in the surgery group was significantly greater: −29.4+/−9.0 kg (gastric bypass) and −25.1+/−8.5 kg (sleeve gastrectomy) versus −5.4+/−8.0 kg with intensive medical intervention. About four times the number of patients in the surgery group achieved the target outcome of HbA1c≤6% at 12 months post-intervention compared to the medical intervention. While initial diabetes improvement/remission has been observed in many studies, this is not sustained in all individuals in the long-term as observed in the Swedish Obese Subjects (SOS) study [26]. Factors associated with diabetes improvement/remission include degree of weight loss, patient age, duration of diabetes, and insulin dependency.

Weight loss is generally considered difficult in patients with type 2 diabetes [29]. A major problem is that many of the older drugs used to treat diabetes (sulphonylureas, thiozolidenediones, and insulin) are associated with weight gain. High endogenous insulin is associated with increased lipid deposition as is administration of exogenous insulin. Furthermore, hypoglycemia is associated with increased food intake, which exacerbates adiposity. Fortunately, novel diabetes treatments [30–32] are associated with lesser degrees of hypoglycaemia and are either weight neutral (dipeptidyl peptidase IV inhibitors, gliptins) or are associated with weight loss (glucagon-like peptide 1 GLP-1 agonists and sodium glucose cotransporter-2 SGLT-2 inhibitors). The weight loss in response to GLP-1 agonists is generally modest, but clinically important at doses used for diabetes control. At higher doses GLP-1 agonist use is associated with greater weight loss [33].

Obesity and Type 2 Diabetes—Shared Complications

Obesity and Type 2 diabetes are part of the cluster of cardio-metabolic disorders that include hypertension, dyslipidemia, non-alcoholic fatty liver disease [34], and obstructive sleep apnoea [35]. Non-alcoholic fatty liver disease consists of a spectrum of disorders with steatohepatitis, fibrosis, cirrhosis, and hepatocellular carcinoma being its worst manifestations. Obstructive sleep apnoea is associated with intermittent complete or partial airway obstruction during sleep accompanied by hypoxemia. Obstructive sleep apnoea is closely related to obesity [36] and is very common in patients with diabetes ranging from about 30–80 % of patients depending on study stetting, patient population studied, and cut-points for diagnosis of sleep apnoea. Both obesity and type 2 diabetes are also associated with several cancers e.g. breast cancer [37, 38], poor mental health and wellbeing [39], cognitive decline [40], and reduced quality of life [39]. Additionally, obese patients with diabetes are at greater risk of diabetes complications. Key pathogenetic mechanisms operating include oxidative stress and inflammation.

Obesity, Type 2 Diabetes, and Mortality

Obesity increases the risk of mortality in the general population, which has been proven in several prospective cohorts from various ethnic backgrounds in different parts of the world [41–44]. However, studies pertaining to mortality in obese patients with diabetes are controversial. While some studies have reported decreased mortality in these patients, a phenomenon known as obesity paradox, others suggested a neutral effect, a U-shaped association or an increased risk only with high BMI ranges (Table 12.1). Many factors contribute to the conflicting results. The studies, to date, have not been congruent in different aspects such as the follow-up period, definition of diabetes at study initiation, the number of participants, and mortality events rates. Most importantly, statistical correction of mortality risk on common CVD risk factors such as smoking has not been universally included.

The mechanisms underlying the increased risk of mortality with significant excess or low adiposity in type 2 diabetes patients are unclear because of association of other confounding co-morbid pathologies. However, the increased risk of mortality is often driven by a significant elevation in cardiovascular and cancer related mortalities. The recent follow-up of approximately 11.500 patients with incident diabetes in 2 large prospective cohort studies showed a significant association with severe obesity and all-cause mortality, cardiovascular mortality and cancer-related mortality (HR 1.33 [95 % CI 1.14–1.55], HR 1.23 [95 % CI 1.01–1.70] and HR 1.33 [95 % CI 1.07–2.00], respectively [45].

Whether weight loss in obese patients with type 2 diabetes is associated with decreased mortality is controversial. An earlier systematic review that included studies of weight loss and mortality in patients with type 2 diabetes concluded that results were contradictory [46]. The meta-analysis of 26 studies that enrolled individuals with BMI >35 Kg/m^2 found out that weight loss is associated with decreased all-cause mortality risk in obese patients that suffer from other co-morbidities such as diabetes [47]. Recently, the 9.6 years follow-up of an intensive lifestyle intervention in overweight or obese patients with type 2 diabetes, the Look AHEAD trial, showed a moderate decrease in CVD risk factors but not death [22].

Table 12.1 Studies reporting association between mortality and obesity in patients with type 2 diabetes

Type of study	Number of patients with type 2 diabetes	Follow-up period	Result	Reference
Pooled analysis of 5 prospective cohorts	2,625	27,125 person-years	Obese patients when diabetes was diagnosed had less risk of death compared to their obese counterparts	[90]
Pooled analysis of 2 prospective cohorts	11,427	15.8 years	J-shaped association between BMI and mortality among non-smokers (BMI range <22.5 or >25 kg/m^2) and a linear association among smokers. Obesity did not confer protection to newly diagnosed patients with diabetes in terms of mortality	[45]
Prospective	106,640	4.7 years	U-shaped association between BMI and mortality. Mortality was increased for BMI range 20 to 25 kg/m^2 or ≥35 kg/m^2	[91]
Prospective	373	14 years	J-shaped association between BMI and mortality. Mortality was increased for BMI range <21.2 or 22.7 and >27.3 or 27.8 kg/m^2 in women and man, respectively	[92]
Prospective	30,534	2.7 years	U-shaped association between BMI and all-cause, total, and CVD mortality for a BMI <25 or >30 kg/m^2	[93]
Prospective	814	8.1 years	U-shaped association between BMI and all-cause mortality for a BMI <25 or >40 kg/m^2	[94]
Prospective	7,534	5.5 years	V shaped association between BMI and all cause-mortality, cut-off BMI being 26 kg/m^2	[95]
Prospective	8,334	8 years	Overweight and/or obesity was not associated with increased risk of mortality	[96]
Prospective	2,960	NA	BMI >26 kg/m^2 was not associated with increased risk of mortality	[97]

Obesity, Type 2 Diabetes, and Cardiovascular Disease

Cardiovascular disease is a major cause of morbidity and mortality in both obesity and diabetes. Diabetes and obesity are associated with dyslipidaemia, hypertension, increased inflammation, oxidative stress, and endothelial dysfunction, which together predispose to cardiovascular disease.

There is no accepted clinical measure that best predicts cardiovascular events in obese patients irrespective of the presence of diabetes. While many clinical studies use standard BMI ranges, it has been suggested that measurement of total body fat and its distribution may be better indicator for estimating the risk of CVD risk in obese patients [48].

During the past decades, cardiovascular complications in patients with diabetes have decreased, notably myocardial infarction and stroke [49]. As a consequence, cause-specific mortality to these major complications has also reduced [50]. Nevertheless, patients with diabetes still have a higher risk of cardiovascular disease compared to the non-diabetic population.

In patients with type 2 diabetes, obesity is not only an additional risk factor that enhances the absolute risk of CVD, but another disease on top of the underlying pathology. In the Framingham heart study, obesity increased the cardiovascular risk in patients with type 2 diabetes by more than 40 % over 30 years, most noticeable in women [51]. Interestingly, the risk in obese patients with type 2 diabetes correlates strongly with insulin resistance as cardiovascular risk profile is higher in patients with high insulin resistance compared to insulin-sensitive individuals in the same BMI range [52]. Additionally, the risk is even enhanced at the pre-diabetes stage as supported by epidemiological data. Analysis from of the Third National Health and Nutrition Examination Survey showed that insulin resistance associated with other components of the metabolic syndrome were independent predictors of myocardial infarction whereas obesity itself was not [53].

Obesity, Type 2 Diabetes, and Coronary Heart Disease

Type 2 diabetes has for some time been recognised as a coronary heart disease (CHD) equivalent [54, 55]. Type 2 diabetes and obesity confer independently a higher risk for fatal and non-fatal CHD events compared to non-diabetic and non-obese individuals [56, 57].

Obesity induces coronary atherosclerosis formation through classical and well-known mechanisms, such as dyslipidemia, hypertension, and other components of the metabolic syndrome. However, the association between obesity and diabetes could enhance the activation of other pathophysiological pathways, such as subclinical inflammation, neurohormonal activation with increased sympathetic tone, increased free fatty acid turnover, and intra-myocardial and sub-epicardial fat deposition [58, 59]. Most importantly, obese patients with type 2 diabetes exhibit disturbances in the coagulation and fibrinolytic pathways, hence platelet adhesion is enhanced and endothelial dysfunction accelerated. High levels of insulin, fibrinogen, plasminogen activator inhibitor-1, factor VII, factor VIII and von Willebrand factor have all been implicated [60, 61]. Although shared genetic susceptibility has been identified for the triple association of obesity-type 2 diabetes-CHD in Genome-Wide Association Studies (GWAS), the genetic predisposition has only a minor impact [62].

Epidemiological and experimental data are not completely aligned. The 6 year follow-up of the Swedish National Diabetes Register suggested that overweight and obese patients with diabetes who were CHD event-free at inclusion were at higher risk for CHD and stroke compared to patients with a BMI <25 kg/m^2 [63]. Larger cohorts or meta-analysis examining only obese patients with type 2 diabetes are lacking. However, an obesity paradox has been proposed for a protective role of obesity in CHD related mortality in studies that included participants with type 2 diabetes. In-hospital mortality and coronary revascularization rates were significantly lower in extremely obese patients, 63.4% of whom had type 2 diabetes, admitted for myocardial infarction compared to those who were not obese [64]. In the National Cardiovascular Data Registry (NCDR, USA), overweight and obesity up to a BMI of 40 kg/m^2 in patients with ST-elevation myocardial infarction was not associated with higher in-hospitality mortality; that was only validated in the group of extreme obese patients of whom 43% were diabetic [65]. In the Swedish Coronary Angiography and Angioplasty Registry, patients with angiographically documented CHD, of whom 41.3% had type 2 diabetes, there was a 3 year increased mortality only when BMI was >35 kg/m^2 [66].

The effect of weight loss on CHD events in obese patients with diabetes is still unclear. In the Look-Ahead trial, coronary events were not reduced in patients with type 2 diabetes who lost weight compared to controls. The Swedish Obese Subjects (SOS) study reported a reduction in the risk of myocardial infarction in obese patients with type 2 diabetes who underwent bariatric surgery over 13 years of follow-up [67]. In the recent 3 year follow-up of a study that compared bariatric surgery to intensive medical therapy, BMI and HbA1c decreased in the surgery group whereas major CHD risk factors such as hypertension and Low-Density Lipoprotein (LDL) were almost unchanged [28].

Obesity, Type 2 Diabetes, and Nephropathy

Both diabetes and obesity are known to increase the risk of renal impairment [68, 69]. However, obese patients with type 2 diabetes are at higher risk of developing diabetic nephropathy (DN) and End Stage Renal Disease (ESRD) [70]. In the Swedish National Population Register, the Odds Ratio (OR) of having diabetic nephropathy was found to be 7.4 (4.2–13.0) for a BMI >35 kg/m^2,and 2.8 (1.8–4.4) for a BMI between 30 and 35 kg/m^2, compared to a BMI <25 kg/m^2 [71]. Similar findings were reported in the follow-up of a large integrated health care delivery system in northern California [72].

Whether the decline of renal function in type 2 diabetes patients with established nephropathy is directly influenced by obesity is controversial with very few studies reporting positive association between obesity and an accelerated decline in renal function (Table 12.2).

The mechanisms mediating the association between obesity and diabetic nephropathy are complex. Excess adipose tissue is believed to secrete a panel of in-

Table 12.2 Effect of obesity on progression of diabetic nephopathy (DN) in patients with type 2 diabetes

Type of study	Number of patients with type 2 diabetes	Follow-up period	Result	Reference
Prospective	229	31 months	High BMI did not influence the rate of progression of DN	[98]
Prospective	152	11 years	BMI was not associated with new onset of DN in patients with normo-albuminuria	[99]
Retrospective	195	4 years	Baseline BMI did not predict the decrease in GFR	[100]
Retrospective	621	9.9 years	BMI was an independent factor for GFR decline in univariate analysis; the association was lost in multivariate analysis	[101]
Prospective	292	5 years	BMI did not predict the progression of ESRD	[102]

flammatory mediators [73] and specific proteins such as adiponectin, leptin and resistin; all playing a pivotal role in the pathogenesis of obesity-associated nephropathy, along with oxidative stress [74]. Additionally, obesity activates the renin-angiotensin aldosterone system that further leads to the progression of kidney failure [75]. Another mechanism includes hypoxemia secondary to obstructive sleep apnoea, which accompanies obesity and type 2 diabetes [35, 76, 77].

Obese patients with type 2 diabetes have insulin resistance and hyperinsulinemia. The latter induces the synthesis of growth factors such as insulin-like growth factor 1 and 2, and TGF-1β, which stimulate production of extra-cellular matrix and accelerate glomerular hypertrophy [78]. Additionally, chronic hyperglycaemia and hyperinsulinaemia in obese patients result in vasodilatation of afferent arterioles, which leads to glomerular hyperfiltration [79]. It is already known that chronic glomerular hyperfiltration is an initial cornerstone in the development and progression of glomerulosclerosis [70, 80, 81]. Biopsy of obese patients with proteinuria, but otherwise healthy, has revealed changes characteristic of either focal glomerulosclerosis or occult diabetic nephropathy, while no changes were seen in normo-albuminuric, non-obese, age and sex controls [82]. Similar histopathological finding were also reported in fatty Zucker rats in whom food restriction decreased glomerulomegaly and mesangial matrix expansion, and limited focal and segmental glomerulosclerosis [83].

The positive correlation between obesity and nephropathy could be generalized to other type 2 diabetes microvascular complications. Obesity is an independent risk factor for diabetic neuropathy; additionally, higher BMI ranges are associated with advanced stages of retinopathy [84]. Finally, weight loss is an efficacious treatment to delay the progression to ESRF and decrease urinary albumin excretion in patients with diabetic nephropathy. This has been proven in different approaches including diet modifications or bariatric surgery [85–87].

Conclusion

There is no indication that the levels of obesity and type 2 diabetes are declining despite public health efforts. Indeed, obesity and diabetes are affecting populations where previously they were rare. Together, these two chronic conditions and their co-morbidities will continue to challenge healthcare services and societies. The predisposing and accelerating biological factors for obesity and diabetes (Fig. 12.2) include: traditional (e.g. lifestyle), increasingly recognised (sleep duration [88, 89], shift-work, chronotype), and novel (epigenetic, gastrointestinal microbiome, endocrine disrupting chemicals) factors. Further understanding of these factors and their underlying molecular mechanisms, and developing strategies to prevent and/or treat their deleterious effects are required to face the combined challenge of obesity and diabetes. Intensive lifestyle interventions, pharmaceutical agents and surgery have all shown their efficacy in treating obese patients with type 2 diabetes, although it is still unknown weather this will improve long-term cardiovascular morbidity and mortality.

References

1. World Obesity. World obesity statistics. http://www.worldobesity.org/aboutobesity/resources/obesity-data-portal/. Accessed July 2014
2. World Health Organization. http://www.who.int/gho/ncd/risk_factors/overweight/en/. Accessed July 2014
3. World Health Organization. http://www.who.int/diabetes/facts/en/. Accessed July 2014
4. Mitchell NS, Catenacci VA, Wyatt HR, Hill JO (2011) Obesity: overview of an epidemic. Psychiatr Clin North Am 34(4):717–732
5. Schmidt M, Johannesdottir SA, Lemeshow S, Lash TL, Ulrichsen SP, Bøtker HE, Sørensen HT (2013) Obesity in young men, and individual and combined risks of type 2 diabetes, cardiovascular morbidity and death before 55 years of age: a Danish 33-year follow-up study. BMJ Open 3(4):e002698
6. Tirosh A, Shai I, Afek A et al (2011) Adolescent BMI trajectory and risk of diabetes versus coronary disease. N Engl J Med 364:1315–1325
7. Henricsson M, Nystrom L, Blohme G et al (2003) The incidence of retinopathy 10 years after diagnosis in young adult people with diabetes: results from the nationwide population-based diabetes incidence study in Sweden (DISS). Diabetes Care 26:349–54
8. Gunathilake W, Song S, Sridharan S, Fernando DJ, Idris I (2010) Cardiovascular and metabolic risk profiles in young and old patients with type 2 diabetes. QJM (monthly journal of the Association of Physicians) 103:881–4
9. Wong J, Molyneaux L, Constantino M, Twigg SM, Yue DK (2008) Timing is everything: age of onset influences long-term retinopathy risk in type 2 diabetes, independent of traditional risk factors. Diabetes care 31:1985–90
10. Bays HE (2011) Adiposopathy is "sick fat" a cardiovascular disease? J Am Coll Cardiol 57:2461–2473
11. Deng Y, Scherer PE (2010) Adipokines as novel biomarkers and regulators of the metabolic syndrome. Ann N Y Acad Sci 1212:e1–e19
12. Cao H (2014) Adipocytokines in obesity and metabolic disease. J Endocrinol 220(2):T47–59

13. Yoo HJ, Choi KM (2014) Adipokines as a novel link between obesity and atherosclerosis. World J Diabetes 5(3):357–363
14. Abi Khalil C (2014) The emerging role of epigenetics in cardiovascular disease. Ther Adv Chronic Dis 5(4):178–187
15. Ruchat SM, Hivert MF, Bouchard L (2013) Epigenetic programming of obesity and diabetes by in utero exposure to gestational diabetes mellitus. Nutr Rev 71(Suppl 1):S88–S94
16. Kahn SE, Cooper ME, Del Prato S (2014) Pathophysiology and treatment of type 2 diabetes: perspectives on the past, present, and future. Lancet 383(9922):1068–1083
17. Alonso-Magdalena P, Quesada I, Nadal A (2011) Endocrine disruptors in the etiology of type 2 diabetes mellitus. Nat Rev Endocrinol 7(6):346–353
18. Moreno-Indias I, Cardona F, Tinahones FJ, Queipo-Ortuño MI (2014) Impact of the gut microbiota on the development of obesity and type 2 diabetes mellitus. Front Microbiol 5:190
19. Colditz GA, Willett WC, Stampfer MJ, Manson JE, Hennekens CH, Arky RA, Speizer FE (1990) Weight as a risk factor for clinical diabetes in women. Am J Epidemiol 132(3):501–513
20. Knowler WC, Barrett-Connor E, Fowler SE et al (2002) Reduction in the incidence of type 2 diabetes with lifestyle intervention or metformin. N Engl J Med 346:393–403
21. Thomas GN, Jiang CQ, Taheri S, Xiao ZH, Tomlinson B, Cheung BM, Lam TH, Barnett AH, Cheng KK (2010) A systematic review of lifestyle modification and glucose intolerance in the prevention of type 2 diabetes. Curr Diabetes Rev 6(6):378–387
22. Wing RR, Reboussin D, Lewis CE, Look AHEAD Research Group (2013) Intensive lifestyle intervention in type 2 diabetes. N Engl J Med 369(24):2358–2359
23. Andrews RC, Cooper AR, Montgomery AA et al (2011) Diet or diet plus physical activity versus usual care in patients with newly diagnosed type 2 diabetes: the early ACTID randomised controlled trial. Lancet 378(9786):129–139
24. Dixon JB, le Roux CW, Rubino F, Zimmet P (2012) Bariatric surgery for type 2 diabetes. Lancet 379(9833):2300–2311
25. Leong WB, Taheri S (2012) The role of bariatric surgery in the treatment of type 2 diabetes mellitus. J R Coll Physicians Edinb 42(3):194–198
26. Sjöström L, Lindroos AK, Peltonen M et al (2004) Lifestyle, diabetes, and cardiovascular risk factors 10 years after bariatric surgery. N Engl J Med 351(26):2683–2693
27. Ribaric G, Buchwald JN, McGlennon TW (2014) Diabetes and weight in comparative studies of bariatric surgery vs conventional medical therapy: a systematic review and meta-analysis. Obes Surg 24(3):437–455
28. Schauer PR, Bhatt DL, Kirwan JP et al (2014) Bariatric surgery versus intensive medical therapy for diabetes—3-year outcomes. N Engl J Med 370:2002–2013
29. Mavian AA, Miller S, Henry RR (2010) Managing type 2 diabetes: balancing HbA1c and body weight. Postgrad Med 122(3):106–117
30. Bohannon N (2009) Overview of the gliptin class (dipeptidyl peptidase-4 inhibitors) in clinical practice. Postgrad Med 121(1):40–45
31. Labuzek K, Kozłowski M, Szkudłapski D, Sikorska P, Kozłowska M, Okopień B (2013) Incretin-based therapies in the treatment of type 2 diabetes–more than meets the eye? Eur J Intern Med 24(3):207–212
32. Fujita Y, Inagaki N (2014) Renal sodium glucose cotransporter 2 inhibitors as a novel therapeutic approach to treatment of type 2 diabetes: Clinical data and mechanism of action. J Diabetes Investig 5(3):265–275
33. Astrup A, Rössner S, Van Gaal L et al (2009) Effects of liraglutide in the treatment of obesity: a randomised, double-blind, placebo-controlled study. The Lancet 374(9701):1606–1616
34. Pallayova M, Taheri S (2014) Non-alcoholic fatty liver disease in obese adults: clinical aspects and current management strategies. Clinical Obesity [in press].
35. Pallayova M, Banerjee D, Taheri S (2014) Novel insights into metabolic sequelae of obstructive sleep apnoea: a link between hypoxic stress and chronic diabetes complications. Diabetes Res Clin Pract 104(2):197–205

36. Young T, Peppard PE, Taheri S (2005) Excess weight and sleep-disordered breathing. J Appl Physiol 99(4):1592–1599
37. Xu CX, Zhu HH, Zhu YM (2014) Diabetes and cancer: associations, mechanisms, and implications for medical practice. World J Diabetes 5(3):372–380
38. Berger NA (2014) Obesity and cancer pathogenesis. Ann N Y Acad Sci 1311:57–76
39. Jagielski AC, Brown A, Hosseini-Araghi M, Thomas GN, Taheri S (2014) The association between adiposity, mental well-being, and quality of life in extreme obesity. PLoS One 9(3):e92859
40. McCrimmon RJ, Ryan CM, Frier BM (2012) Diabetes and cognitive dysfunction. Lancet 379(9833):2291–2299
41. Adams KF, Schatzkin A, Harris TB et al (2006) Overweight, obesity, and mortality in a large prospective cohort of persons 50 to 71 years old. N Engl J Med 355:763–778
42. Berrington de Gonzalez A, Hartge P, Cerhan JR et al (2010) Body-mass index and mortality among 1.46 million white adults. N Engl J Med 363:2211–19
43. Zheng W, McLerran DF, Rolland B et al (2011) Association between body-mass index and risk of death in more than 1 million Asians. N Engl J Med 364:719–729
44. Chen Z, Yang G, Offer A et al (2012) Body mass index and mortality in China: a 15-year prospective study of 220,000 men. Int J Epidemiol 41:472–481
45. Tobias DK, Pan A, Jackson CL et al (2014) Body-mass index and mortality among adults with incident type 2 diabetes. N Engl J Med 370:233–244
46. Williamson DF (1998) Weight loss and mortality in persons with type-2 diabetes mellitus: a review of the epidemiological evidence. Exp Clin Endocrinol Diabetes Off J Ger Soc Endocrinol Ger Diabetes Assoc 106(Suppl 2):14–21
47. Harrington M, Gibson S, Cottrell RC (2009) A review and meta-analysis of the effect of weight loss on all-cause mortality risk. Nutr Res Rev 22:93–108
48. Poirier P (2009) Cardiologists and abdominal obesity: lost in translation? Heart Br Card Soc 95:1033–1035
49. Gregg EW, Li Y, Wang J et al (2014) Changes in diabetes-related complications in the United States, 1990–2010. N Engl J Med 370:1514–1523
50. Abi Khalil C, Roussel R, Mohammedi K et al (2012) Cause-specific mortality in diabetes: recent changes in trend mortality. Eur J Prev Cardiol 19:374–81
51. Fox CS, Pencina MJ, Wilson PWF et al (2008) Lifetime risk of cardiovascular disease among individuals with and without diabetes stratified by obesity status in the Framingham heart study. Diabetes Care 31:1582–1584
52. McLaughlin T, Abbasi F, Lamendola C, Reaven G (2007) Heterogeneity in the prevalence of risk factors for cardiovascular disease and type 2 diabetes mellitus in obese individuals: effect of differences in insulin sensitivity. Arch Intern Med 167:642–648
53. Ninomiya JK, L'Italien G, Criqui MH et al (2004) Association of the metabolic syndrome with history of myocardial infarction and stroke in the third national health and nutrition examination survey. Circulation 109:42–46
54. Beckman JA, Paneni F, Cosentino F, Creager MA (2013) Diabetes and vascular disease: pathophysiology, clinical consequences, and medical therapy: part II. Eur Heart J 34:2444–2452
55. Paneni F, Beckman JA, Creager MA, Cosentino F (2013) Diabetes and vascular disease: pathophysiology, clinical consequences, and medical therapy: part I. Eur Heart J 34:2436–2443
56. Turner RC, Millns H, Neil HA et al (1998) Risk factors for coronary artery disease in non-insulin dependent diabetes mellitus: United Kingdom Prospective Diabetes Study (UKPDS: 23). BMJ 316:823–828
57. Global Burden of Metabolic Risk Factors for Chronic Diseases Collaboration (BMI Mediated Effects), Lu Y, Hajifathalian K et al (2014) Metabolic mediators of the effects of body-mass index, overweight, and obesity on coronary heart disease and stroke: a pooled analysis of 97 prospective cohorts with 1.8 million participants. Lancet 383:970–983

58. Poirier P, Giles TD, Bray GA et al (2006) Obesity and cardiovascular disease: pathophysiology, evaluation, and effect of weight loss: an update of the 1997 American heart association scientific statement on obesity and heart disease from the obesity committee of the council on nutrition, physical activity, and metabolism. Circulation 113:898–918
59. Chaowalit N, Lopez-Jimenez F (2008) Epicardial adipose tissue: friendly companion or hazardous neighbour for adjacent coronary arteries? Eur Heart J 29:695–697
60. Alessi MC, Peiretti F, Morange P et al (1997) Production of plasminogen activator inhibitor 1 by human adipose tissue: possible link between visceral fat accumulation and vascular disease. Diabetes 46:860–867
61. Eliasson M, Røder ME, Dinesen B et al (1997) Proinsulin, intact insulin, and fibrinolytic variables and fibrinogen in healthy subjects. A population study. Diabetes Care 20:1252–1255
62. Wu C, Gong Y, Yuan J et al (2012) Identification of shared genetic susceptibility locus for coronary artery disease, type 2 diabetes and obesity: a meta-analysis of genome-wide studies. Cardiovasc Diabetol 11:68, (another publication: Diabetes Care. 2008;31(9):1847–1852
63. Eeg-Olofsson K, Cederholm J, Nilsson PM et al (2009) Risk of cardiovascular disease and mortality in overweight and obese patients with type 2 diabetes: an observational study in 13,087 patients. Diabetologia 52:65–73
64. Dhoot J, Tariq S, Erande A et al (2013) Effect of morbid obesity on in-hospital mortality and coronary revascularization outcomes after acute myocardial infarction in the United States. Am J Cardiol 111:1104–1110
65. Das SR, Alexander KP, Chen AY et al (2011) Impact of body weight and extreme obesity on the presentation, treatment, and in-hospital outcomes of 50,149 patients with ST-Segment elevation myocardial infarction results from the NCDR (National Cardiovascular Data Registry). J Am Coll Cardiol 58:2642–2650
66. Angeras O, Albertsson P, Karason K et al (2013) Evidence for obesity paradox in patients with acute coronary syndromes: a report from the Swedish coronary angiography and angioplasty registry. Eur Heart J 34:345–353
67. Romeo S, Maglio C, Burza MA et al (2012) Cardiovascular events after bariatric surgery in obese subjects with type 2 diabetes. Diabetes Care 35:2613–2617
68. Marre M, Lievre M, Chatellier G et al (2004) Effects of low dose ramipril on cardiovascular and renal outcomes in patients with type 2 diabetes and raised excretion of urinary albumin: randomised, double blind, placebo controlled trial (the DIABHYCAR study). BMJ 328:495
69. Wang Y, Chen X, Song Y et al (2008) Association between obesity and kidney disease: a systematic review and meta-analysis. Kidney Int 73:19–33
70. Ting SMS, Nair H, Ching I, Taheri S, Dasgupta I (2009) Overweight, obesity and chronic kidney disease. Nephron Clin Pract 112(3):c121–c127
71. Ejerblad E, Fored CM, Lindblad P et al (2006) Obesity and risk for chronic renal failure. J Am Soc Nephrol JASN 17:1695–1702
72. Hsu C, McCulloch CE, Iribarren C et al (2006) Body mass index and risk for end-stage renal disease. Ann Intern Med 144:21–28
73. Tang J, Yan H, Zhuang S (2012) Inflammation and oxidative stress in obesity-related glomerulopathy. Int J Nephrol 2012: 608397
74. De Ferranti S, Mozaffarian D (2008) The perfect storm: obesity, adipocyte dysfunction, and metabolic consequences. Clin Chem 54:945–955
75. Bosma RJ, van der Heide JJH, Oosterop EJ et al (2004) Body mass index is associated with altered renal hemodynamics in non-obese healthy subjects. Kidney Int 65:259–265
76. Banerjee D, Leong WB, Arora T, Nolen M, Punamiya V, Grunstein R, Taheri S (2013) The potential association between obstructive sleep apnea and diabetic retinopathy in severe obesity-the role of hypoxemia. PLoS One 8(11):e79521
77. Leong W, Banerjee D, Nolen M, Adab P, Neil Thomas G, Taheri S (2014) Hypoxemia and glycemic control in type 2 diabetes mellitus with extreme obesity. J Clin Endocrinol Metab 99(9):E1650–E1654. (jc20141260)

78. Aihara K-I, Ikeda Y, Yagi S et al (2010) Transforming growth factor-β1 as a common target molecule for development of cardiovascular diseases, renal insufficiency and metabolic syndrome. Cardiol Res Pract 2011:175381
79. Juncos LA, Ito S (1993) Disparate effects of insulin on isolated rabbit afferent and efferent arterioles. J Clin Invest 92:1981–1985
80. Brenner BM (1983) Hemodynamically mediated glomerular injury and the progressive nature of kidney disease. Kidney Int 23:647–655
81. Abi Khalil C, Travert F, Fetita S et al (2010) Fetal exposure to maternal type 1 diabetes is associated with renal dysfunction at adult age. Diabetes 59:2631–36
82. Kasiske BL, Crosson JT (1986) Renal disease in patients with massive obesity. Arch Intern Med 146:1105–1109
83. Maddox DA, Alavi FK, Santella RN, Zawada ET Jr (2002) Prevention of obesity-linked renal disease: age-dependent effects of dietary food restriction. Kidney Int 62:208–219
84. Dirani M, Xie J, Fenwick E et al (2011) Are obesity and anthropometry risk factors for diabetic retinopathy? The diabetes management project. Invest Ophthalmol Vis Sci 52:4416–4421
85. Navaneethan SD, Yehnert H, Moustarah F et al (2009) Weight loss interventions in chronic kidney disease: a systematic review and meta-analysis. Clin J Am Soc Nephrol CJASN 4:1565–1574
86. Afshinnia F, Wilt TJ, Duval S et al (2010) Weight loss and proteinuria: systematic review of clinical trials and comparative cohorts. Nephrol Dial Transplant Off Publ Eur Dial Transpl Assoc—Eur Ren Assoc 25:1173–1183
87. Jose B, Ali A, Begum S, Banerjee D, Taheri S (2011) Obstructive sleep apnoea as a cause of headache presenting to the emergency department. QJM 104(12):1087–1089
88. Taheri S (2007) Sleep and metabolism: bringing pieces of the jigsaw together. Sleep Med Rev 11(3):159–162
89. Taheri S, Mignot E (2010) Sleep well and stay slim: dream or reality? Ann Intern Med 153(7):475–476
90. Carnethon MR, De Chavez PJD, Biggs ML et al (2012) Association of weight status with mortality in adults with incident diabetes. JAMA 308:581–590
91. Logue J, Walker JJ, Leese G et al (2013) Association between BMI measured within a year after diagnosis of type 2 diabetes and mortality. Diabetes Care 36:887–893
92. Ross C, Langer RD, Barrett-Connor E (1997) Given diabetes, is fat better than thin? Diabetes Care 20:650–652
93. Khalangot M, Tronko M, Kravchenko V et al (2009) Body mass index and the risk of total and cardiovascular mortality among patients with type 2 diabetes: a large prospective study in Ukraine. Heart Br Card Soc 95:454–460
94. Hanson RL, McCance DR, Jacobsson LT et al (1995) The U-shaped association between body mass index and mortality: relationship with weight gain in a Native American population. J Clin Epidemiol 48:903–916
95. So WY, Yang X, Ma RCW et al (2008) Risk factors in V-shaped risk associations with all-cause mortality in type 2 diabetes-The Hong Kong diabetes registry. Diabetes Metab Res Rev 24:238–246
96. McEwen LN, Karter AJ, Waitzfelder BE et al (2012) Predictors of mortality over 8 years in type 2 diabetic patients: translating research into action for diabetes (TRIAD). Diabetes Care 35:1301–1309
97. Chaturvedi N, Fuller JH (1995) Mortality risk by body weight and weight change in people with NIDDM. The WHO multinational study of vascular disease in diabetes. Diabetes Care 18:766–774
98. Mohsen A, Brown R, Hoefield R et al (2012) Body mass index has no effect on rate of progression of chronic kidney disease in subjects with type 2 diabetes mellitus. J Nephrol 25:384–393
99. Viswanathan V, Tilak P, Kumpatla S (2012) Risk factors associated with the development of overt nephropathy in type 2 diabetes patients: a 12 years observational study. Indian J Med Res 136:46–53

100. Pavkov ME, Knowler WC, Lemley KV et al (2012) Early renal function decline in type 2 diabetes. Clin J Am Soc Nephrol 7:78–84
101. Alwakeel JS, Isnani AC, Alsuwaida A et al (2011) Factors affecting the progression of diabetic nephropathy and its complications: a single-center experience in Saudi Arabia. Ann Saudi Med 31:236–242
102. Bentata Y, Latrech H, Abouqal R (2014) Does body mass index influence the decline of glomerular filtration rate in diabetic type 2 patients with diabetic nephropathy in a developing country? Ren Fail 36(6):838–846

MIX
Papier aus verantwortungsvollen Quellen
Paper from responsible sources
FSC® C105338

If you have any concerns about our products,
you can contact us on
ProductSafety@springernature.com

In case Publisher is established outside the EU,
the EU authorized representative is:
Springer Nature Customer Service Center GmbH
Europaplatz 3, 69115 Heidelberg, Germany

Printed by Libri Plureos GmbH
in Hamburg, Germany